EXPERIENCE
TECHNOLOGY
MANUFACTURING
CONSTRUCTION

Henry R. Harms
Technology Education Teacher
Smithtown Central School District
Smithtown, NY

Dennis K. Kroon, Ed.D.
Technology Education Department Chair
Hauppauge Public Schools
Hauppauge, NY

Marlene Weigel
Technical Writer
Peoria, IL

GLENCOE

Macmillan/McGraw-Hill

New York, New York
Columbus, Ohio
Mission Hills, California
Peoria, Illinois

Send all inquiries to:
GLENCOE DIVISION
Macmillan/McGraw-Hill
3008 W. Willow Knolls Drive
Peoria, IL 61614

ISBN: 0-02-646946-4 (Text)
ISBN: 0-02-646953-7 (Teacher's Resource Binder)
ISBN: 0-02-646956-1 (Student Workbook)

Printed in the United States of America

    3 4 5 6 7 8 9 10 RRW 96 95 94 93

Acknowledgments

The publisher wishes to thank the following people, who served as planning consultants for this book.

Mr. Joe Charles "Chuck" Bridge
Industrial Technology Education Teacher
Chisholm Trail Middle School
Round Rock, TX

Mr. Ronnie McQueen
Technology Teacher
Madison Middle School
Abilene, TX

Ms. Barbara Brock
Industrial Technology Teacher
Friendswood, TX

Mr. Alan L. Towler
Technology Education Curriculum Specialist
The University of Texas at Austin
Austin, TX

For his help in planning and developing this program, special thanks to

Dr. Ronald E. Jones
Professor and Coordinator of Technology Teacher Education
Department of Industrial Technology
University of North Texas
Denton, TX

Table of Contents

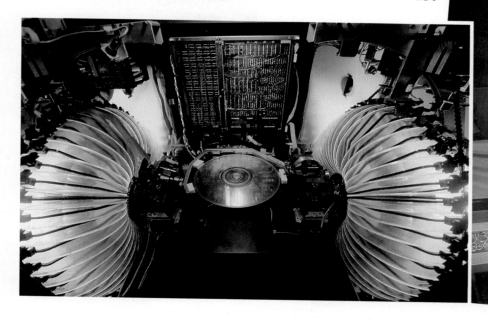

Getting Acquainted with This Book

You have probably heard it said that experience is the best teacher. This means that we learn best by actually doing something. The title of this textbook is *Experience Technology: Manufacturing and Construction*. Notice the first word in the title — "experience." In this course, you will learn through experience — by "doing." You will not only read, you will also talk with classmates during discussions. You will work with them on many interesting activities.

The next word in the title is "technology." What is technology? Technology is the use of the principles of science to do something. For example, workers on a construction site use scientific principles when they build a house.

The last words in the title are "manufacturing" and "construction." In this course, you will see how technology applies to both.

How This Book Is Organized

This book is divided into five sections. Sections I and II cover the things that manufacturing and construction have in common. Section III discusses just manufacturing, and Section IV covers construction. Then in Section V the two are brought together again and their future is discussed.

Section Activities

In most books, the activities are at the end of a section. You have to study the material before you begin to work. This book is different. Each section opens with an activity that you can get started on right away. The activity is described in a brief that outlines the situation and your challenge. Specifications, limits, and a materials list are given. Also included is a list of references you can go to for help. Pictures, charts, and other aids make the activities more interesting and easy to work with.

Chapters

These come after the section activity. The chapters are easy to read and have a lot of pictures. Some of the pictures fill one or two pages and contain interesting information that you will want to know about.

Section Closings

Each section closes with a feature called "You Can Make a Difference." These are true stories about young people who have used technology to invent things or to make the world a better place.

What You'll Find in the Chapters

There are 18 chapters in this book.

Chapter Openings

Each chapter begins with an introduction, a list of objectives, and a list of words you will need. The introduction tells you what the chapter is about.

The objectives state what you should be able to do after you've studied the chapter. The "Words you will need" are terms used in technology. Understanding these will help you understand the chapter. These terms can be found in **boldfaced type** within the chapter. They are also defined in the Glossary.

The ways in which technology affects the world and society are its impacts. For example, the "Impact" for Chapter three talks about how we have become part of a larger world because of technology. At least one impact of technology is included in each chapter.

Discussion Questions

Each major part of a chapter is followed by one or more discussion questions. These help you think about what you have read by discussing the answers with your classmates.

Company Activities

These appear after the discussion questions. Your teacher will probably lead you in forming your own student company. These activities relate to that company.

Extension Activities

Extension activities are suggestions of things you can do to apply and expand what you've learned.

Chapter Review

Chapter review pages include three sections. "Chapter Highlights" list important points to remember from the chapter. "Test Your Knowledge" includes 10 questions that ask you to remember key facts in the chapter. "Correlations" are activities that show how technology can be applied in science, math, language arts, and social studies.

Glossary, Index, and Appendix

At the end of the book are three sections that can be helpful to you as you study — the glossary, the index, and the appendix.

Glossary

If you need to know the meaning of a term, look in the glossary. Terms listed at the beginning of each chapter are also listed in alphabetical order in the glossary. You'll find other important words there too.

Index

If you would like to know where a certain subject is discussed in this book, look in the index. It will give you the numbers of pages that provide information on that subject.

Appendix

There are two appendices. Appendix A is titled, "Measuring with the Metric System." It will help you when you want to design in metric or convert customary measures to metric ones. Appendix B is called, "How to Conduct a Meeting." It talks about the rules used by groups when holding meetings. It may help you if you form a student company.

INTRODUCTION TO PRODUCTION SYSTEMS

Activity Brief
Developing a Company Philosophy

PART 1 : Here's the Situation...........

In Section 1, you will get to know production systems. You will learn how they are organized, what they do for us, and how they affect our lives. You will discover what is involved in forming a company. You will also find out why a company chooses to sell one product rather than another.

While you are learning these things, your teacher will lead you in forming your own student company. Your company will manufacture and/or build a product that you will then sell or use in some way. This activity will help you in forming that company. It will get you started in thinking about the kind of company it should be.

PART **2**: Your Challenge...........

Every company has a philosophy. A philosophy is a set of thoughts and beliefs. The people who create the company are usually responsible for its philosophy. They have their own ideas about what is important to them and how a business should be run.

For example, Patagonia, Inc., is a real company based in California. It makes outdoor clothing. Its owner believes in making unique, high-quality products. He is also concerned about the environment. His beliefs about these things affect the way the company is run. Since quality is so important, only the best materials are used. Since the environment is also important, the company gives at least 10 percent of its profits each year to saving the earth. You will read more about Patagonia on these pages.

For this activity, you will do three things. Together with your classmates, you will determine the philosophy of your student company and put it into written form. Then you will discuss the effect this philosophy will have on the way you conduct your business. Finally, you will design a package for your product that communicates your company's philosophy.

PART **3**: Specifications and Limits...........

Your work will need to meet certain standards. Read the following specifications and limits before you begin.
1. Your company's philosophy must come to grips with at least three questions:
   • What are your company's most important goals?

Patagonia, Inc. got its start when its founder began to make rock climbing equipment.

   • What is your company's responsibility to its customers?
   • What is your company's responsibility to the community?
2. The way your package is designed and *made* must reflect the company's philosophy. For example, suppose your company believes in conserving natural resources. Would some kinds of packaging be wasteful of these resources? Could you make do with less? You might then want to consider a bag, tag, or label of some kind as your "package."

3. You must hand in the following:

- a paragraph describing how your package reflects the company's philosophy
- a log sheet of your work
- rough sketches of at least 3 designs you considered
- a finished drawing of the design you like best
- a mock-up of the package, including any message that would be printed on it; the complete message may be written on a separate sheet of paper

PART 4: Materials...........

There are many materials you might use to make a package. Here is a list of basics.

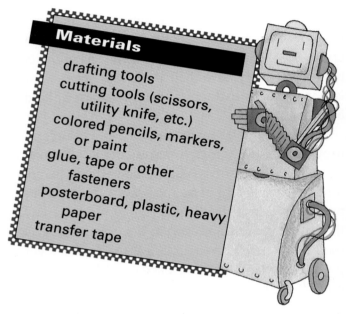

Materials

drafting tools

cutting tools (scissors, utility knife, etc.)

colored pencils, markers, or paint

glue, tape or other fasteners

posterboard, plastic, heavy paper

transfer tape

PART 5: Procedures...........

How you work will be up to you. Still, there are certain steps to follow that may make your work easier.

1. Each day, keep a log of the work you do. Remember, this log must be turned in with your finished project.

2. Determine your company's philosophy before you do anything else. You may want to work in small groups at first. Then, when you have some good ideas, discuss them as a class. Read "What Makes Patagonia Special?" on page 17. Try to focus on things you really believe in rather than things you *think* you should believe in. Put your own values into the company.

3. When you're satisfied with your philosophy, vote on it as a class. When your philosophy has passed a majority vote, post it in the technology lab.

4. Study your company's product. How should it look in the package? Should the package hold more than one? What materials would work best?

5. Make several sketches of possible packages. Then choose the design you like best. Read "What Do Customers Want?" on page 15 and "What's Your Advertising Strategy?" on page 16.

6. Make a careful drawing of the design, including all measurements. Submit this final drawing to your teacher for approval before going on.

7. Make a mock-up (model) of the package. On the package or on a separate sheet of paper, write the message the package should carry. Be sure your company's philosophy is somehow communicated.

What Do Customers Want?

Customers have made the following complaints about some companies or products. How will your company address these problems?

- Companies don't care about customers. They don't give anything back to the community. All they care about is making money.
- Certain products are hurting the environment. Others waste energy.
- Too often, products are of poor value. They do not hold up well under use. They are not guaranteed.
- Companies do not provide enough information about how to use and care for their products. When instructions exist, they are unclear.
- Companies don't want to hear about complaints. Their phone numbers are often unlisted.
- Prices on products are too high.
- Packages are misleading. They make a product look bigger or better than it really is.
- Companies don't care about their employees. They don't care if workers are satisfied or happy. They don't let them get involved in making decisions.

PART 6: For Additional Help...........

With this activity is information that may be of use to you as you determine a philosophy and make a package. Additional information about the following subjects is found on these text pages:

advertising a product, pp. 191-193
customer needs and wants, p. 21
economy, effects of production on, pp. 29-33
environment, effects of production on, pp. 28-29
forming a company, pp. 36-51
market research, pp. 54-58
packaging, pp. 186-190
society, effects of production on, p. 33

Patagonia specializes in clothing for all kinds of weather.

What's Your Advertising Strategy?

An advertising strategy is a plan for persuading customers to buy your product. It is used when designing labels and packages. Many companies ask themselves the following questions. Then they create their packages around the answers.

- How will a customer benefit from doing business with us?
- What needs does a customer have that our product can fulfill?
- How can we show customers that it is safe to do business with us?

Research Tip:
Look up the meanings of these words:
ethical
commitment
responsible

PART 7: How Well Did You Meet the Challenge?..........

When you've finished, evaluate your work. Does your package do what it's supposed to do? Your teacher and classmates may take part in this evaluation.

1. How well does your package or communicate the company philosophy?

2. How well does your package display or protect the product?

3. How could your package be improved?

PART 8: Extending Your Experience..........

Think about the following questions and discuss them in class.

1. What kinds of effects, or impacts, do you think your company's philosophy will have on customers and on your school?

2. What kinds of impacts did your company's philosophy have on your own work? What might you have done differently if you hadn't decided on a philosophy first?

3. Name several well-known companies. From what you can tell by the way these companies do business, what would you say their philosophies are?

What Makes Patagonia Special?

The founder of Patagonia, Yvon Chouinard (left), likes to go rock climbing. As a young man, he was dissatisfied with the quality of climbing equipment. He decided he could make better pitons and carabiners, two items that climbers use. (The steel spike shown here is a piton. Climbers hammer them into cracks in the rock and use them to climb on. The metal rings are carabiners. They snap into pitons. A rope is then threaded through them.) Chouinard bought some blacksmith's equipment and set up a forge in his parents' backyard. He sold his pitons to friends.

People liked his equipment. Soon he and a friend began redesigning other climbing equipment. They formed a company to sell it. Their equipment became the most popular on the market. Then they discovered there was a need for good outdoor clothing, too. Chouinard insisted that the quality of the clothes be as high as that of the pitons and carabiners. He found

fabrics made with advanced technology. He used clothing designs that fit well and held up under hard wear. Before long, the Patagonia brand was the most popular outdoor clothing made. Even astronauts on Shuttle missions have worn Patagonia's special underwear under their flight suits!

Because Yvon Chouinard is concerned about the environment, Patagonia donates money each year to saving the earth. Recently, he has also decided to limit his company's growth. This is because making certain products uses limited natural resources, such as oil. To conserve these resources, Patagonia will make only a certain number of products per year, and no more. However, the products will be the very best the company can produce. They will last the wearer for many years.

Stores that carry Patagonia clothes say that they outsell comparable brands by 5-to-1. Yet the company does little advertising. Some clothing companies put out 22 mail-order catalogs a year. Patagonia publishes only two. What do you think makes this company so successful?

CHAPTER **1**

Getting to Know Production Systems

Introduction......................................

When you woke up this morning, you were lying in a bed created by a production system. The toothbrush, toothpaste, soap, and towels you used were also results of production systems. So were the foods you ate for breakfast and the car or bus you took to school. The school itself was built by a production system, and so was the house or apartment you'll go home to tonight. Without production systems, your life would be very different.

After studying this chapter, you should be able to......................

Describe production systems.

Tell the difference between needs and wants and give examples of how production systems try to satisfy them.

Use the universal systems model to understand different systems.

List the six steps in problem solving and use them to solve problems.

Describe several impacts of production systems.

Words you will need ....

construction
consumers
feedback
input
manufacturing
output
process
production systems

productivity
profit
recycling
resource
system
technology
universal systems model

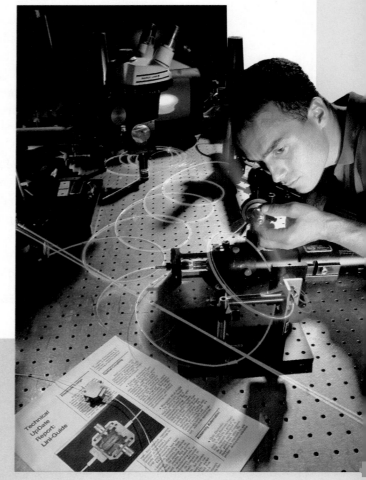

What Is a System?

A **system** is an organized way of doing something. Your school, for example, is part of the educational system. It is an organized way of providing education. Some systems already exist in nature. The solar system is an example. Other systems, such as production systems, are created by people. A **production system** is an organized way of manufacturing or constructing a product. Fig. 1-1.

Some large systems are composed of smaller systems called subsystems. Fig. 1-2. A computer system has subsystems — the keyboard, central processing unit, disk drive, and so on. Production systems, too, have subsystems. You will learn about some of them in this course.

Fig. 1-1. Both manufacturing and construction are production systems. They are organized ways of making a product.

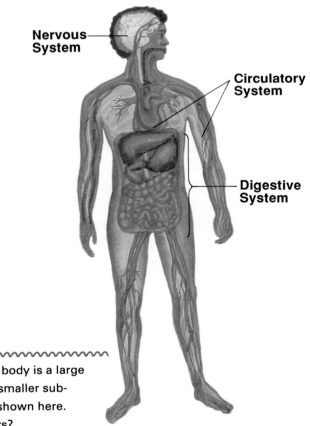

Fig. 1-2. The human body is a large system made up of smaller subsystems. Three are shown here. Can you name others?

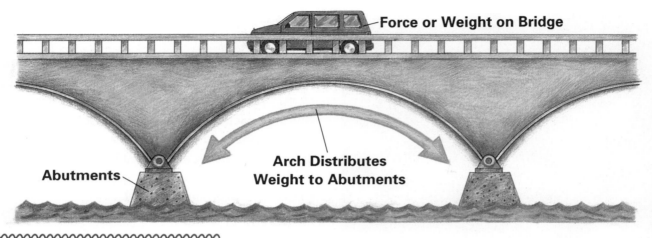

Force or Weight on Bridge

Arch Distributes Weight to Abutments

Abutments

Fig. 1-3. This bridge design is based on scientific principles. If it were not, it might easily collapse.

Systems Use Technology

Production systems are technological systems. People create **technology** when they apply the principles of science to do something. Workers in production systems depend on scientific principles to make products. For example, when engineers design a bridge, they make use of scientific principles regarding force and loads. Fig. 1-3. The bridge must withstand the force of moving water and the weight of traffic. For purposes of this course, we can say that technology is the use of knowledge to make products that meet people's needs and wants. Today, science and technology are related fields of study.

All the goods and structures we use are created by production systems using technology. Transportation, communication, energy/power, and biotechnology are also technological systems.

Systems Have Goals

The parts of a system work together to achieve a goal. For example, the goal of the educational system is to produce educated human beings. Production systems have basically two goals. The first is to

manufacture or construct a product. Products made by means of **manufacturing** are created in a factory or some other processing facility. Bread, cakes, and even ice cream cones, for example, are produced in a bakery. Fig. 1-4. **Construction** uses manufactured items to build something *on site* — the place where it will be used. A house is built on the lot where it will be located. Fig. 1-5. Lumber, windows, doors,

Fig. 1-4. Many of our processed foods, such as these ice cream cones, are made by food manufacturers.

Fig. **1-5**. The site for this house is the lot on which it is built.

and other manufactured products are brought to the site and installed.

The second goal of a production system is to sell the product and make a profit. **Profit** is the money left over after all the bills and workers are paid. A successful production system will usually be more profitable.

Fulfilling Needs and Wants

In order to fulfill their goal of selling products, production systems must make products that people need or want.

We need certain things to keep us alive and well. Food is a need, as are clothes and shelter. Wants are those things that make our lives easier or more pleasant. A new compact disc player and a diamond ring are both wants. All these things are provided by production systems.

Often, the society we live in determines some of our needs. For example, electrical power is considered a need in the United States. Fig. 1-6. Without electricity some of us would freeze in winter. Our

food would quickly spoil. We would not have power to run our computers.

Sometimes the same item may be a want *or* a need, depending on the situation. For a person who lives in a town having public transportation, a car may not be necessary. In that case, a car would be considered a want. What if the person lives in the country, far from bus lines? Then a car may be essential. It becomes a need.

Fig. **1-6**. How many electrical appliances can you find in this picture? Without power, they are useless.

▶▶▶ FOR DISCUSSION ◀◀◀

1. **Some systems are natural. They already exist in nature. The solar system is one example. Can you think of others?**

2. **Electrical power is a need in the United States. Describe a country where it might not be a need.**

3. **Tell whether the following are wants or needs and give your reasons: a glass of milk, a bicycle, a television set.**

Company Activity

■ **If your class forms a production company, it will produce a product. Make a list of at least three products your class could produce. Tell whether each is a want or a need and why. Next, indicate why you think someone would want to buy the product.**

How to Understand Systems

All systems operate in basically the same way. They can be easily understood using the universal systems model.

The Universal Systems Model

The **universal systems model** is a diagram that shows how any system works. Fig. 1-7. Every system has input, a process, and output. Many systems also have feedback.

Input

Input includes anything that is put into a system. The input comes from resources. A **resource** is anything that provides supplies or support for the system. All technological systems require input from seven basic resources. Fig. 1-7.

- people
- information
- materials
- tools and machines
- energy
- capital (money)
- time

Let's apply the universal systems model to a system you're familiar with. Consider a school technology laboratory's cleanup system. The *goal* of the system is to have a clean lab at the end of each class period. Here are the inputs for this system.

People: The teacher and students.

Information: The explanation by the teacher of what needs to be done and how to do it.

Materials: The things that were used during the class period that need to be put away.

Tools and machines: Brooms, dustpans, vacuum cleaner, waste containers.

Energy: Both human energy and the electrical power used to run machines.

Capital: Money from the school used to purchase the materials, tools, and machines.

Time: The time required to get the job done.

The Process

The **process** is the action part of a system. It is what *happens* to the inputs or resources. Inputs are combined by means of two main actions: managing and producing. Fig. 1-7.

Managing involves planning, organizing, and controlling what is done. Producing is the actual making of the product.

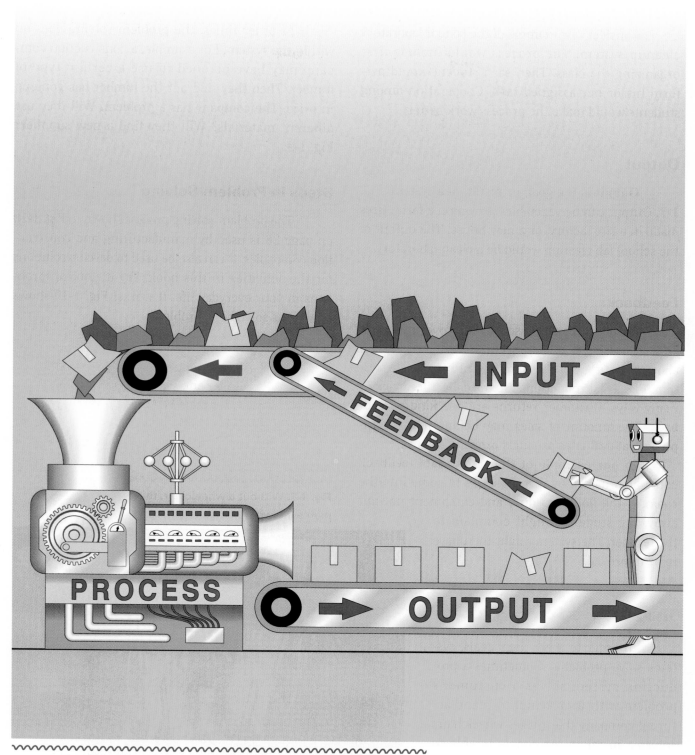

Fig. 1-7. Resources are put into the system. There they are processed. Products are the output of the system. Feedback is what is learned about the products and sent back to the system.

Consider our example of the school laboratory cleanup system. The process would include first organizing the class. Then each student would perform his or her assigned task. Cooperation among students would make the process work better.

Output

Output is the goal, or result, of a system. Fig. 1-7. Output can be vegetables grown on a farm, tires made in a tire factory, or a new house. The output of the school lab cleanup would be a clean laboratory.

Feedback

Feedback can be used to improve the way a system works. **Feedback** occurs when information about the output of a system is sent back to the system. Fig. 1-7. Production systems collect feedback in many ways. Customer returns are one kind of feedback. The amount of sales may be another. A computer that evaluates products provides a third.

In our example of the school lab, feedback could be provided by the next class to come into the room. If the floor was dirty and equipment not put away, the students might complain to the teacher. The system might then have to be improved.

The Problem-Solving Process

In order to meet their goals, systems solve problems. Sometimes a production system solves a customer's problem with its product. A manufacturing company that makes wheelchairs solves a problem for people unable to walk. Fig. 1-8. A construction company may solve housing problems by building more new homes.

At other times, the problem solving goes on within the system. For example, a construction company may have planned to use a certain type of lumber. Then they find out the lumber has gone up in price. The company has a problem. Will they use different materials? Will they find a new supplier? Fig. 1-9.

Steps in Problem Solving

The problem-solving process shown in Fig. 1-10 on page 26 is used by manufacturing and construction companies. It can also be used to develop solutions for the activities in this book. Try it, too, on problems in your everyday life. It works! Fig. 1-10 shows how teens solved one problem.

Fig. 1-8. Without a wheelchair, this man's life would be quite different.

Fig. 1-9. This construction boss has to decide how to solve the problem of a price increase. Here he is gathering information about work already done.

IMPACT

Competition is an impact of production systems. Competition among manufacturers has inspired them to develop new materials, techniques, and products. We have all benefited from these innovations. Fifteen years ago, no one had ever heard of a VCR (video cassette recorder). Today, millions of families who own VCRs can enjoy movies at home.

Manufacturing companies use brainstorming to come up with new product ideas, product names, or advertising slogans. During this course, you will be doing some brainstorming. You'll discover just how helpful it can be.

▶▶▶ FOR DISCUSSION ◀◀◀

1. Use the universal systems model to describe how your school cafeteria operates.
2. Sean is failing in math because he has no place to study. Using the six steps in problem solving, suggest ways in which Sean could solve his problem.

Brainstorming

One way to get ideas for problem solving is by brainstorming. During brainstorming, a group of people suggests ideas. All the ideas are recorded by one member. All ideas, no matter how wild, are listened to and written down. Even though a wild idea may not be usable itself, it can inspire a usable idea. After a period of time, the ideas are discussed and evaluated. Some are thrown away. Others may be combined or changed in some way.

Company Activity

■ As a class, brainstorm to come up with product ideas for your student company. (You may suggest those you used for the activity on page 22.)

Fig. 1-10. Carrying heavy schoolbooks is a problem for these teens. Using the problem-solving process, they come up with a solution.

THESE BOOKS WEIGH A TON!

The problem is stated clearly.
Before a problem can be solved, it must be understood. Questions, such as *who, what, where, when, why,* and *how* are asked.

Information is collected.
What needs to be known to develop a solution? Sources of information include libraries, museums, computerized data bases, and people who are experts on the subject.

Several solutions are developed.
Most problems have more than one solution. As many as possible should be considered.

The best solution is selected.
The advantages and disadvantages of each suggested solution are considered. There is usually no perfect solution to a problem. Most solutions have their good points and their bad points. A compromise is usually required. The solution that looks the most promising is selected.

GREAT BOOK BAGS!

GREAT JOB!

The solution is put to work.
The solution is tried in an actual situation to be sure it works as planned. Some solutions succeed or fail immediately. Others may take a little longer before results are known.

The solution is evaluated.
The solution may need adjustments. Sometimes a solution must be scrapped completely. In that case, steps 4, 5, and 6 are repeated with other solutions.

Impacts of Production Systems

Production systems are linked to one another and to other systems. They are also an important part of our daily lives. They have an impact, or effect, on almost everything we do. They affect the environment, the economy, and our society as a whole.

IMPACT

Social trends have an impact on construction. During the 1970s and 80s, many people moved to the "Sun Belt" — the southern and southwestern states. Construction in those areas boomed. The Midwest lost population, and there the construction industry slumped.

Links to Other Systems

Manufacturing and construction systems could not exist without one another. Fig. 1-11. Without factories built by construction systems, manufacturers could not make products. Without manufactured items, like tools and lumber, construction companies could not build structures.

Production systems are also linked to other technological systems in many ways. Those include transportation, communication, energy/power, and biotechnical systems.

Transportation systems rely on manufactured and constructed products. Every airplane is manufactured. Every airport is constructed. Production systems, in turn, also rely on transportation systems. Without roads or trucks, manufactured items could not be shipped to buyers.

Fig. 1-11. This hospital could not be built without manufactured products. How many can you find in the picture?

Fig. **1-12**. This worker at the right is making circuit boards that fit into a computer. The finished computer is soon part of the communication system these workers below need to set up schedules.

machine is another. Biotechnology depends on production systems to manufacture the medicines and machines. Production systems depend on biotechnology to keep human workers healthy and able to do their jobs.

Effects on the Environment

The environment is everything that surrounds us. It includes the air, land, water, and all living things. Production systems have damaged the environment. Acid rain, solid waste, and water pollution are some of the problems. At the same time, production systems are working toward solving some of those problems.

Acid Rain

When fuels are burned, they produce sulfur and nitrogen oxides. In the air, these combine with water to produce sulfuric acid. The result is called acid rain. When acid rain falls, it damages trees and lakes. Fig. 1-13. Fish and other water life have died because of acid rain.

In communication systems, every printing press, telephone, and computer has been manufactured. Fig. 1-12, Part A. Without production systems, we would not be able to communicate as well as we do now. Without communication, production systems could not use computers to run machines. Orders for products could not be called in by telephone. Fig. 1-12, Part B.

Energy/power systems depend on production systems to construct the dams and power plants that create electricity. Production systems depend on energy/power systems to provide the electricity to run machines.

Biotechnical systems use knowledge about biology and engineering to solve human problems. Medicines are one result of biotechnology. The X-ray

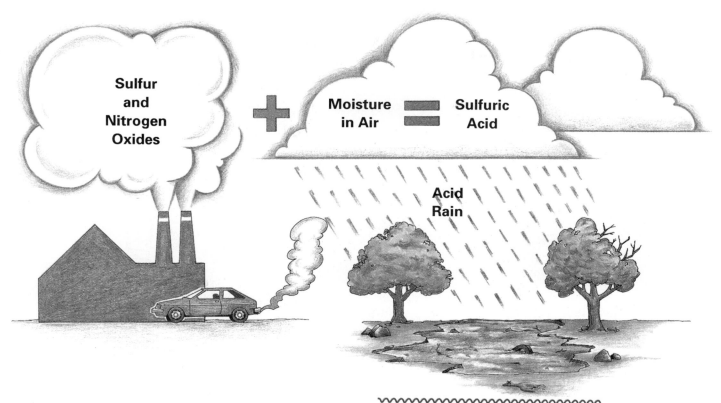

Fig. 1-13. Acid rain does serious damage to the environment.

Solid Waste

What happens to the things we throw away? Many communities have problems getting rid of solid waste. Glass, metal, and paper products are dumped into landfills. Landfill space, however, is running out. Some states have encouraged recycling of waste products. **Recycling** means reusing the materials. When recycling is done, there is less litter. Metals can be reclaimed. Fig. 1-14. Fewer trees are cut down to make paper. Less garbage needs to be burned or buried.

Water Pollution

Production systems can pollute rivers, lakes, and groundwater supplies. (Groundwater is underground water from which we get water to drink.) Federal and state laws now require that industries purify the water they have used. Fig. 1-15. More and more water is being recycled.

Effects on the Economy

Economics is the study of how goods and services are produced, distributed, and used. The economic system used in the United States is called free enterprise, or capitalism. Under this system, most goods and services are produced by privately owned companies. **Consumers** buy the goods and services. We are all consumers.

When the economy is healthy, people spend money to buy products. The number of production jobs increases. When the economy is slow, there are fewer jobs. People don't have as much money to spend. Production systems are an important part of the economy.

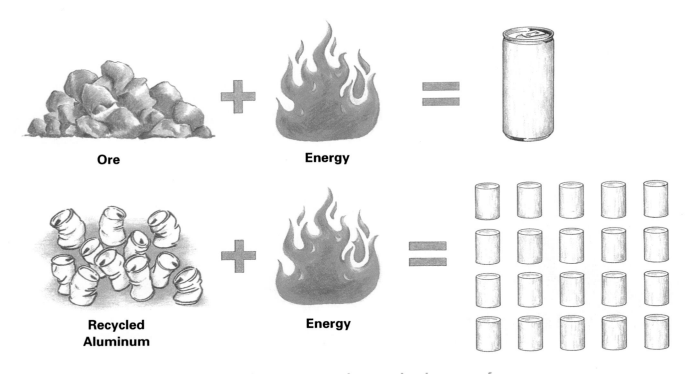

Ore   Energy

Recycled   Energy
Aluminum

It takes as much energy to make <u>one</u> aluminum can from new material as it does to make <u>20</u> cans from recycled material.

Fig. 1-14. Metals can be reused. This saves both the metal and the energy used to refine it.

IMPACT

An increase in the number of manufactured products is good for the national economy. When manufacturers have profitable years, they pay more taxes. These taxes provide for many government services. Taxes also inspire more growth. The greatest impact is usually on the local or state economy. New jobs are created. New homes may be built for workers moving into the area.

Fig. 1-15. Polluted rivers destroy wildlife.

In 1990, about 15 percent of the population was employed in manufacturing jobs. This number is decreasing for two reasons. First, new techniques have made workers more productive. Fewer workers are needed to produce the same amount of goods. Second, more and more companies are providing services rather than making products. Fig. 1-16. Services are in great demand in our country. The United States is the first nation to have a large share of its workers *not* producing food or products.

Construction, too, has an important effect on the economy. Building supplies, plumbing fixtures, and banking services must be provided for construction projects. This creates work for other industries.

Production systems also help establish our country's productivity. **Productivity** is a measure of how many resources (inputs) it takes to make a certain number of products (output). Let's say company B makes twice as many products with the same number of resources as company A. Company B has higher productivity than company A. Fig. 1-17.

Fig. 1-16. Limousine services have become popular in recent years. They often serve as shuttles from and to airports so that travelers may leave their cars at home. They also provide elegant transportation for those going to weddings and proms.

Fig. 1-17. Which company, A or B, do you think makes a bigger profit?

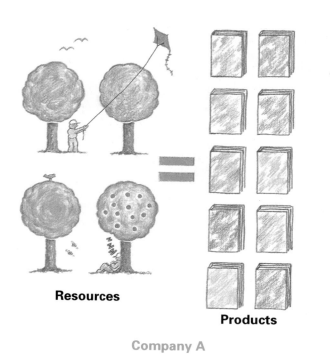

Resources

Products

Company A

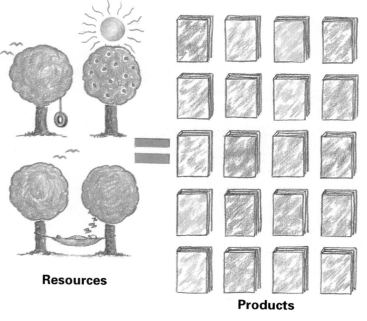

Resources

Products

Company B

Impacts of the Automobile

The first automobile insurance was issued by Traveler's Insurance Company in 1897.

In 1895, air-filled tires began to be used on cars. Many manufacturers now provide other items for car maintenance, such as air filters, batteries, and antifreeze. What others can you name?

Were it not for the automobile, drive-up windows would not exist. Banks, fast-food restaurants, and cleaners have all felt the impact.

DRIVE-UP WINDOW

Adam's
HOME & AUTO
INSURANCE

Traffic lights, signs, and other equipment are needed for traffic control.

Cars have created jobs. Highway and traffic patrols are needed to regulate car traffic.

GAS

Businesses that service automobiles form an industry of their own. Body repair shops, "minute" lubrication centers, and detail shops are only a few. Can you name others?

The first parking meter was used in Oklahoma City, Oklahoma, in 1935. What positive and negative impacts have resulted from parking regulations?

NO PARKING

Of the world's top 10 manufacturing firms, 7 are oil companies.

Fig. 1-18. Can you think of other impacts the automobile has had on our lives?

Technology can increase productivity. For example, computers can help factories make more products in less time.

People can also improve productivity. In Japan, South Korea, and West Germany, workers are very productive. They have good work attitudes. Their jobs are very important to them. In some industries they are more productive than American workers. This is a serious problem because the United States is less able to compete in world markets.

Effects on Society

New products can change our lives. Before the 1920s, most people still depended on horses for transportation. Then the automobile became very popular and changed everything. New jobs were created in the auto industry. Other workers, who made horseshoes or wagon wheels, were no longer needed. The production of gasoline became a major industry. Gas stations popped up everywhere. Traffic laws, signs, and signals had to be created. Accidents occurred and sparked a new service — car insurance. None of these things would have happened had it not been for the automobile. Fig. 1-18.

What about computers and robots? Less than 25 years ago, few people knew much about computers. Now they are everywhere. Manufacturing companies use them to make products. Architects use them to design buildings. More and more people have a computer in their homes. Computerized robots are replacing assembly line workers. Robots are used to weld products, spray them with paint, and put them in storage. Fig. 1-19.

Production methods have helped create new kinds of health problems. People who need to keep up with high-speed computers and other new machines report more job stress. Carpal tunnel syndrome, a painful wrist condition, has affected some people who use computers or supermarket scanners.

As production systems change, they will affect us in many ways. Some effects will be good and others will be bad. We must work together to lessen the bad effects and share the good ones. All through this book you will find short features about impacts of production systems.

▶▶▶ FOR DISCUSSION ◀◀◀

1. Imagine that a new restaurant is about to open in your neighborhood. What systems will have been linked in making this happen? What did each system provide?
2. Are there any industries in your town responsible for air or water pollution? What are they and how do they harm the environment?
3. Have any companies gone out of business recently in your town? Has this affected people you know? How?

Extension Activity

■ Using a camera, photograph an environmental problem in your community. Create a poster with the photos. Write captions for them. Label ways in which production systems may have helped cause the problem shown. Tell what should be done about the problem.

Fig. 1-19. In this General Motors plant, some robots pick up items and place them elsewhere. Others weld parts, and still others apply paint.

Chapter Highlights

● A system is an organized way of doing something. The two production systems are manufacturing and construction.

● The goal of a production system is to make products. The products fulfill wants or needs.

● The universal systems model describes all systems. All systems have inputs, a process, and outputs. Many have feedback.

● Systems solve problems. Brainstorming is a way of collecting new ideas for problem solving.

● Production systems influence the world around us. They are linked to other systems. They affect the environment, the economy, and our society.

Test Your Knowledge

1. Tell what a system is and give an example.

2. Name the two production systems and tell how they differ.

3. Name the four parts of the universal systems model.

4. Name the seven resources used as inputs to production systems.

5. List in order the six steps in problem solving.

6. What is the term used for a group of people coming up with ideas and writing them down?

7. What four other technological systems are often linked to production systems?

8. What is produced when sulfur and nitrogen oxides mix with water in the atmosphere?

9. What is the name of the economic system used in the United States?

10. What is the term used for the measure of how many resources it took to make a certain number of products?

Correlations

LANGUAGE ARTS

1. Write a business letter to the Environmental Protection Agency. Ask for a copy of their rules and regulations for solid waste disposal.

SOCIAL STUDIES

1. As a class compare the needs and wants that a 19th-century American had with the needs and wants that a modern American has.

SCIENCE

1. Obtain samples of water from lakes, ponds, rivers and streams in your area. Test how acidic they are by using universal indicator paper.

MATH

1. Check the 1990 census for the population of your community. If 15% of the people were employed in manufacturing jobs in 1990, how many would that be?

CHAPTER **2**

Forming a Company

Introduction.....................................

Many companies grow around the idea for a product. The product may be something no one has ever seen before. Other companies may get their start when someone thinks there is a better way to make a product already on the market. Companies may also be started by people who simply want independence. They have worked for other companies and have some experience. They use that experience to create their own businesses.

In this chapter you will learn how a production company is formed. A company's two most basic resources, money and people, will also be discussed. The other resources will be covered in Section II.

After studying this chapter, you should be able to......................

Describe the types of companies and a company's plan of action.

Discuss possible sources of capital and the kinds of expenses a company may have.

Define cash flow and tell why it's important.

Name the kinds of workers a company may use and tell how human resources are managed.

Words you will need ....

**affirmative action
    program**
bonds
capital
cash flow
collective bargaining
corporation
dividends
entrepreneur

expenses
**human resource
    management**
interest
lease
loan
shares
stocks

What Kind of Company Will It Be?

Before a company can hire workers and begin production, it must make some decisions. What kind of company is it to be, and how will it operate?

Types of Companies

Most companies are small to begin with. They may be organized by one person or a group of people. A person who organizes a company is called an **entrepreneur**. That person is responsible for the company and any risks involved.

There are three types of companies based on ownership: proprietorship, partnership, and corporation. Fig. 2-1.

Proprietorship

When a company is owned by one person, it is called a proprietorship. The owner is called the proprietor. Many small businesses are proprietorships. They include farms, beauty salons, and appliance repair shops.

Because proprietorships are small, they need less money to start out with. The owner makes all the decisions. The decisions and any changes can be made quickly. The owner keeps all the profits.

Proprietorships also have disadvantages. Running a company requires many talents. Few people have all the talents that are needed. Often it is hard for the owner to raise extra money when needed. The owner is also responsible for all debts. If the company fails — and many small businesses do — the owner must pay any debts with personal funds.

Partnership

If two or more people own a company, it is called a partnership. The partners share in the decision-making and in the risks. There is more talent available. The owners have more ways to raise money if needed. Any profit is shared.

Fig. 2-1. Owners may be one person, a few, or a large number.

Proprietor   **Partners**   **Corporation**

Corporation

Corporations are companies owned by a number of people. Corporations usually start as proprietorships or partnerships. In order to have money to grow larger, the owners incorporate. As a corporation, the company can legally act as though it were a single person. It can take out loans and sell property. The owners then sell portions of the company, called **shares**. Fig. 2-2. Shares are sold on a stock exchange like the one in Fig. 2-3.

If the company is successful, the shareholders can share in the profits. The profits may be paid in portions called **dividends**.

If the company fails, the shareholders do not have to pay the company's debts. Buildings and equipment are sold instead. However, the money used to buy the shares is lost.

Corporations have some disadvantages. The shareholders are usually not the managers. They often have little interest in how the company is run. Their main interests are in dividends and having their shares go up in price.

A Plan of Action

Every company needs a plan of action. The plan includes the kind of product to be made and the resources needed. For example, how much money is needed to start the company? Where will the money come from? What jobs will have to be done and who will do them?

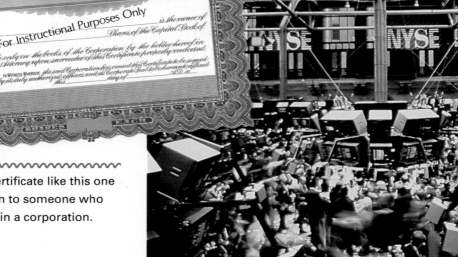

Fig. 2-3. Shares of stock can be bought and sold at a stock exchange.

Fig. 2-2. A certificate like this one may be given to someone who buys shares in a corporation.

The plan also includes how the work force will be organized. Who will be the president of the company? What departments should the company be divided into? For example, most manufacturing companies have four main departments. They include personnel, research and development, production, and marketing. (These departments will be discussed in detail in a later chapter.)

Every company decides on certain policies, or rules, to keep things running smoothly. These rules let workers know what is expected of them. They include such things as working hours and safety rules.

After all these decisions have been made, the company can begin to gather its resources. Perhaps the two most important resources are money and people. Without them the other resources cannot be purchased or used.

Extension Activity

■ **Suppose you want to form your own company. You will do yard work in your neighborhood during the summer. Prepare a plan of action for that company. Include a list of the seven basic resources. Tell what type and quantity of each you will need. Then indicate where you will get them. For example, your list of equipment will probably include a lawn mower. If it is a power mower, under "Energy" be sure to list gasoline. Under "Capital" indicate how much the fuel will cost and where the money will come from to pay for it.**

▶▶▶ FOR DISCUSSION ◀◀◀

1. **What talents do you think are required to run a company? Give examples.**

2. **If you were to start your own company, would you want it to be a proprietorship or a partnership? Explain your reasons.**

3. **The term "division of labor" means that the work is divided into a number of tasks. Different people do different tasks rather than each person trying to do them all. Most companies operate using this principle. What do you think its advantages and disadvantages would be?**

What About Money?

Money is required to start a company and keep it running. This money is called **capital**. It is used to build the factories, buy the equipment and supplies, and pay the workers. Capital may also be in the form of buildings, equipment, or goods. How a company manages its capital affects its financial success.

Sources of Capital

Capital can come from many sources. It may be in the form of loans or material goods. Fig. 2-4.

Where Capital Comes From

PERSONAL SAVINGS

Many small companies are started with the personal savings of the founder. Entrepeneurs often risk their own funds when starting out.

BANK

Banks make loans only to businesses that look like they will be successful. They want to know a company's plan for success. They also want to be sure the loan will be repaid.

PRIVATE INVESTORS

Sometimes family members, friends, and others make a loan. A **loan** is money that will be repaid along with interest. **Interest** is a fee paid for the use of the money.

SMALL BUSINESS

ADMINISTRATION

The Small Business Administration (SBA) is a federal agency that gives loans to small businesses. Like other lenders, the SBA must decide if they think the company will be successful.

BUSINESS

STOCKS AND BONDS

Corporations may decide to sell more shares, also called **stock**, to the public. They may also sell bonds. **Bonds** represent loans made for a certain interest rate and time period. When the bond is due, the company must pay the money back.

SUPPLIERS

Suppliers and equipment manufacturers lend goods or equipment to a company. Payment is made each month toward purchase of these items. The company owns the items after the loan is paid in full.

PROFITS

Sometimes profits are used to expand a company, develop new products, or buy equipment. When this happens, investors may not receive dividends.

Fig. 2-4. All of these may be sources of capital for a company.

Expenses

Capital is needed to pay expenses. **Expenses** are the money needed to run a company and make the product. In general, profits grow when expenses are reduced. Companies have many kinds of expenses. Only the major ones will be discussed here.

Workers

Workers expect to be paid for what they do. Some are salaried. This means they make a certain amount for the year. The money is divided and given to them in weekly, biweekly, or monthly amounts. Managers are usually salaried.

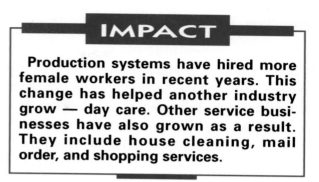

IMPACT

Production systems have hired more female workers in recent years. This change has helped another industry grow — day care. Other service businesses have also grown as a result. They include house cleaning, mail order, and shopping services.

Other workers are paid by the hour. How much money they earn over a year depends on how many hours they work. Most factory workers are paid by the hour.

Worker expenses also include benefits. Among the benefits a company provides are insurance plans, vacations, and a pension. Fig. 2-5.

The government requires that companies contribute to certain benefits. An employee may be injured on the job and unable to work. His or her salary and medical expenses are then paid for out of a fund set aside for that purpose. Workers who lose their jobs may also receive money called unemployment compensation. This money helps them pay living expenses while they look for a new job.

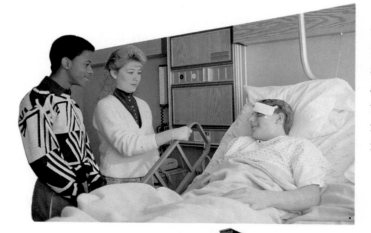

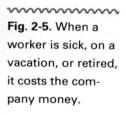

Fig. 2-5. When a worker is sick, on a vacation, or retired, it costs the company money.

Materials, Tools, and Machines

Companies need materials, tools, and machines to make products. Materials are purchased from suppliers. Tools and machines are bought from equipment manufacturers. A company can also lease equipment. A **lease** is a long-term rental agreement. The company pays a monthly fee. After the lease period is over, the equipment must be returned. Leasing allows a company to use equipment without needing a lot of capital to buy it. Equipment that is subject to rapid changes in technology is often leased.

Utilities

Utilities are services provided by a community. The services include such things as energy, water, and communications.

Energy is needed to operate a production company. It may be in the form of electricity for lights and equipment. It may also be in the form of fuel, such as gas for a furnace.

Water is needed for drinking, certain production processes, and sanitation. Telephone service allows workers to talk to one another and to suppliers. All these services are part of a company's expenses.

Taxes

Taxes are collected by local, state, or federal governments. Land, buildings, and other property are usually taxed by the town or state. Federal and state governments tax company profits. The company must also pay Social Security taxes for each worker it employs.

Debts

If the company has borrowed capital, it has debts. A debt is money that is owed. The company usually has to pay a certain portion of a debt each month.

Dividends

If the company has sold stock, it often pays dividends to the stockholders. Each share of stock earns an equal amount. Dividends are usually paid four times a year.

Cash Flow

Money is constantly moving through a company. **Cash flow** is the comparison of money going out to money coming in. A positive cash flow means more money is coming in than is spent. A positive cash flow means there is a profit. Fig. 2-6, Part A.

A negative cash flow means that more money is being spent than is coming in. The company is operating at a loss. Fig. 2-6, Part B. A production system cannot operate for long with a negative cash flow. However, some companies may have a negative flow for certain months and a positive flow for others. What matters is that the company shows a profit at the end of the year.

▶▶▶ **FOR DISCUSSION** ◀◀◀

1. Some companies have reduced expenses by cutting employee benefits, such as insurance. Do you think this is a good idea or not? Why?

2. If you were asked to invest in a production company, what questions would you want answered first?

3. What kinds of expenses do you think your school has? How does the school pay for these expenses? How might expenses be reduced?

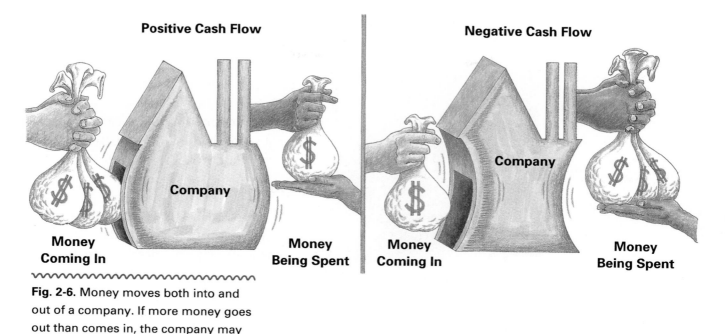

Positive Cash Flow

Company

Money Coming In

Money Being Spent

Negative Cash Flow

Company

Money Coming In

Money Being Spent

Fig. 2-6. Money moves both into and out of a company. If more money goes out than comes in, the company may not survive.

Extension Activity

■ **Review the stock market reports in a newspaper. Pick out a manufacturing or construction stock and imagine that you have bought 200 shares. How much did you pay for each share? How much all together? Follow this stock's prices for one month, keeping records. How much money did you make or lose on the investment?**

Who Will Do the Work?

Like capital, people are essential to a company. Whether the company hires only one employee or one thousand, nothing will get done without them. Some workers make the products. Others come up with ideas. Still others organize tasks.

Kinds of Workers

Basically, there are four types of workers in a production company. They are managers, professionals, technicians, and production workers.

Managers

Managers are people who organize the work. They divide it up among the different workers. Each worker must be told what needs to be done and how to do it. A good manager helps workers perform well. They are made to feel an important part of the production team. Fig. 2-7.

Planning is an important part of management. A plan helps the company achieve its goals. Managers plan how many products to make, when and where to make them, and how much to charge. Information is a key resource used for planning.

Scheduling, too, is done by managers. Lumber must be at the site before carpenters can frame a house. Parts must arrive at a factory before assembly workers can put them together. Good managers see to it that materials are on hand and on time.

Fig. 2-7. In this company, good workers receive awards.

Managers also supervise the work. They are in control of production activities. To be sure the system operates as planned, a manager may move workers around. A manager may also choose to buy from different suppliers or change the rate of production.

Large companies may have many managers at many different levels. Small companies may have only a few. In a small company each manager may have many roles to perform. Most companies use organizational charts to show levels of management and responsibilities. Fig. 2-8.

Professionals

Professionals are those workers having at least a four-year college degree. Some have advanced degrees. Production companies especially need two types of professionals — engineers and architects.

Engineering is the second largest profession in the United States. Engineers use science and math to solve production problems. In both manufacturing and construction, engineers develop methods for making products. They need to know the properties of materials to be used. For example, if concrete is used in a bridge, engineers need to know where to reinforce it with steel rods. Fig. 2-9.

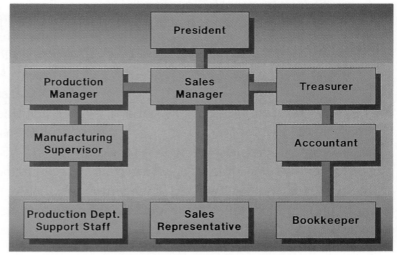

Fig. 2-8. The organization chart shows areas of responsibility.

Fig. 2-9. Engineers apply their knowledge of math and science to solve problems.

Architects plan, design, and supervise all kinds of construction. They talk to clients to learn their needs. Plans are prepared by the architect and approved by the client. During construction, the architect works closely with the building crew. This assures that plans are followed.

Technicians

Technicians have special skills. A two-year college program trains them to build and maintain equipment.

They often assist engineers. In industry, electronics technicians use computers to learn the best way to make a product. They are also responsible for maintaining automated production equipment. Fig. 2-10.

Production systems are becoming more complex. More and more technicians will be needed to see that they operate properly. This means that more jobs for technicians will be available in the future.

Production Workers

In manufacturing, production workers are involved in processing materials. They change materials into products by means of cutting, welding, and other methods. Such jobs are called unskilled, semi-skilled, or skilled, depending on their difficulty.

The largest group of skilled workers in this country is in construction. It includes structural, finishing, and mechanical workers. Structural workers include carpenters, bricklayers, and iron workers. Finishing workers do such things as painting, roofing, and floor installation. Workers in mechanical trades include plumbers and electricians.

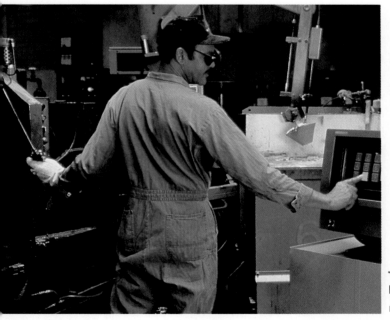

Fig. 2-10. This technician is programming the computer that runs a robot.

IMPACT

Construction projects create many jobs. Architects and engineers plan construction projects. Many different workers are employed during building. An office building, for example, requires surveyors, excavators, masons, carpenters, electricians, and plumbers. Maintenance of the completed building creates additional jobs that exist during the lifetime of the structure.

Managing Human Resources

Almost all companies have a personnel department. This department is concerned with the company's employees. The tasks carried out by this department are called **human resource management**. These tasks include recruiting, hiring, training, retraining, advancing, and separating workers.

Recruiting

After a company decides what jobs must be done, qualified workers must be found. Some jobs are advertised in newspapers or on signs outside the building. For these, people often walk in and fill out an application.

To fill professional jobs, the company sends people to visit colleges and interview students. Fig. 2-11. Professionals may answer ads the company places in newspapers and journals. A professional sends the company a resumé. The resumé gives information about the person's education and work experience.

Hiring

The human resources department holds interviews. Interviews give applicants and employers a chance to get acquainted. Both parties try to learn if they would be happy working together.

Most companies contact an applicant's former employers. They want to know if the person did a good job. They ask about work habits and skills. Sometimes applicants are tested to measure special skills. These might include math ability or skill in using one's hands. The best-qualified applicant is then offered the job.

By law, companies cannot refuse to hire an applicant because of race, religion, sex, marital status, or age. This is so everyone will have an equal opportunity to work. Many companies have **affirmative action programs**. These programs encourage minorities to apply for jobs.

Training

New employees are usually told about the company and benefits. Benefits include such things as sick leave, vacations, and health insurance.

Fig. 2-11. These company representatives are talking to students about jobs with their companies. Students with the right backgrounds may be offered jobs before they graduate.

Employees may also need training. There are several types of training programs.

On-the-job-training. Some employees are trained as they do the work. The training is done by the supervisor or an experienced worker.

Apprenticeships. An apprentice goes to school and also learns on the job. Fig. 2-12. In many industries labor unions sponsor apprenticeship programs. These attract new workers to the field. Also, this helps the unions get new members.

Co-op Programs. A co-op program combines part-time work and part-time education. In this way students can gain work experience while they're in school. Many students are offered a permanent job with the firm later.

Fig. 2-12. This apprentice is learning about car repair from an experienced worker. He learns while he earns.

Retraining

Because of changes in technology, jobs change. New skills may be needed. Retraining programs are used to upgrade worker skills. Workers become qualified for different or better jobs. The company's productivity is increased.

Advancing

When workers advance, they go on to higher-level jobs. The workers get more responsibility and more pay. Under equal opportunity laws, every worker must have the same chance to advance. Many companies help workers advance by sharing the cost of more schooling.

Separating

When a worker leaves the company, he or she is separated. Separating can be permanent or temporary. It may be the employee's decision or the company's.

Layoffs are temporary separations. They occur in companies with money problems. For example, bad weather and a poor economy can cause construction layoffs. Sometimes the workers are recalled after a while. In other cases they are never recalled.

An employee may choose to leave the company for another job elsewhere. This separation is usually permanent. Sometimes workers are fired. Being fired is also permanent. Fired workers are usually not rehired. Workers may be fired for many reasons. They include being late, being absent too much, and not doing a good job.

When workers retire, they no longer work for the company but still receive benefits. The main benefit is called a pension. A pension is money put aside over the years by both the worker and the company. Pensions are based on former salary and number of years with the company. Labor unions also provide benefits.

The Company Team

Employers and employees work together as a team. Each has responsibilities to the other. Some are determined by law. Others may come from the company or labor unions.

Worker Responsibilities

As part of the team, employees must do their assigned jobs to the best of their ability. They must be on time, be honest, and have a good attitude. Employees who help others and cooperate to get the job done are prized. Fig. 2-13.

Attendance is important, too. If a worker is absent, the job may go undone. Someone else may have to do it. An absent employee can harm productivity.

Workers also share the responsibility for safety. Most accidents are caused by carelessness. A worker might use a broken ladder or forget to put a guard on a machine. Employees must follow directions for using equipment. They must use safety gear and follow safety rules. They should also report unsafe conditions to a supervisor.

Employer Responsibilities

Employers must obey equal opportunity laws. They must pay workers regularly for the jobs they do. They must also provide a safe and healthy place to work.

Occupational illnesses are those caused by workplace hazards. Hazards may include such things as dangerous materials or high noise levels. Fig. 2-14. Accidents and illnesses are costly to both the company and the worker.

Many companies have safety programs. These programs teach and promote safety. Companies that use them have fewer accidents. Almost all employees must follow the rules of the Occupational Safety and Health Administration (OSHA). Fig. 2-15. OSHA is part of the United States Department of Labor. Health and safety standards are set and enforced by OSHA.

Fig. 2-13. Workers who help one another also help the company. The job gets done faster and better.

Fig. 2-14. Some factories have high noise levels. Workers' hearing may be damaged. Workers must wear ear plugs or other protection.

Fig. 2-15. These workers are posting a list of safety rules recommended by OSHA.

Labor Relations

Employers and employees together usually make work rules. Sometimes the rules are in the form of a written contract. Contracts are usually made during a process called collective bargaining.

Collective bargaining occurs when a group of workers bargains with the company for wages, benefits, and work conditions. The workers are usually members of a labor union. The union represents the workers during talks with the company. Both sides discuss their concerns and try to come to an agreement. When an agreement is reached, the terms are spelled out in the contract.

Fig. 2-16. When companies and unions cannot agree, a strike may be called. Striking workers picket the company and try to prevent others from entering.

▶▶▶ **FOR DISCUSSION** ◀◀◀

1. How do you think a manager might use the problem-solving process? Give an example for each step of the process.
2. Suppose you have two job offers. What factors would you consider before choosing between the two companies?
3. Have you ever known a worker who was on strike? What was the effect of the strike on the person? Do you think strikes are effective weapons?

———————

If employees feel that the company is not keeping its part of the bargain, they may file a complaint. Then, one of two things usually happens. A third party may be called in to make recommendations. If both sides agree to them, the problem is settled.

When no agreement is reached, a strike or lockout may be called. Fig. 2-16. If employees vote to strike, they refuse to work and are not paid. In a lockout, the company closes the plant. Employees cannot work and are not paid. Both actions are taken to force the other side to bargain. Both actions are costly. Production and sales go down and paychecks stop.

Company Activity

■ Your student company will need to hire workers. One important qualification for every worker is the ability to do the job in question. What other qualifications will you look for? Make a list of all the qualities you think your workers should have. Consider including such things as honesty and neatness. Give the reasons why you think each quality is important to the success of your company. When you have finished, discuss your lists in class. Then create an advertisement for the workers you need. List the qualities the class agrees are important.

Chapter Highlights

● There are three different types of companies — proprietorships, partnerships and corporations. Each one uses a plan of action to get things done.

● Companies require capital, usually in the form of money, to start up and keep running.

● Capital comes from various sources, such as banks and private investors.

● All companies have many expenses which affect cash flow. Cash flow compares the money going out to money coming in.

● The kinds of workers needed by production systems include managers, professionals, technicians, and production workers.

● The human resources department handles all matters concerning the company's employees.

● Employers and workers reach agreements about work conditions by collective bargaining.

Test Your Knowledge

1. Name the three types of companies and briefly describe each.

2. What is a plan of action?

3. Define capital and list five sources of capital.

4. Why are debts considered an expense?

5. What is a negative cash flow?

6. Who are the two main professionals used by a production company and what does each do?

7. Describe what managers do.

8. What are the main duties of human resources departments?

9. What are some training programs available to new employees?

10. Define collective bargaining.

Correlations

LANGUAGE ARTS

1. Locate the word "capital" in a dictionary. Copy the definition for the word, including its part of speech. Which entry or entries apply to the way in which the term is used in this chapter?

SOCIAL STUDIES

1. Research the history of the Massachusetts Bay Company and the American Telephone and Telegraph Company. What individual(s) contributed to the creation of these companies? How did their organization differ?

SCIENCE

1. Interview an engineer or architect. Find out how the person uses science in his or her job.

MATH

1. Suppose your company borrowed $10,000 from a bank at 8% interest for 3 years. How much interest will the company pay to the bank altogether? (Interest = Principal × Rate × Time)

CHAPTER 3

Researching and Developing a Product

Introduction.....................................

Almost everyone has heard of the Apple Computer Company. It is an example of a company that developed around a product. Two entrepreneurs believed that people would want to buy their computers. They started a company in order to manufacture them. What about a company that has already been formed and needs new products to sell? How does this company get ideas for products?

Both construction and manufacturing companies do research. Their research helps them learn about new methods and materials. They also discover what customers want or need.

After studying this chapter, you should be able to.......................

Tell the difference between basic and applied research.

Describe methods used in market research.

Discuss how products are designed.

Tell what needs assessments are and how building sites are chosen.

Describe the process used to design a structure and make estimates for it.

Words you will need ....

applied research

basic research

estimate

market research

patent

prototype

research and
    development

site

specifications

supply and demand

test marketing

working drawings

zoning laws

Kinds of Research

Production systems use three kinds of research. Basic and applied research have to do with the products themselves. Market research has to do with customers.

Basic Research

When you do research, you explore. **Basic**, or pure, **research** is done to learn new information. Scientists and engineers often do this kind of research. Fig. 3-1. They experiment with different materials to find out what happens to them. Also, they try new methods for making and doing things. Sometimes they make exciting discoveries.

Applied Research

Applied research, or development, aims to solve a problem. Scientists and engineers take the information gained in fundamental research and find ways to use it. For example, when NASA decided to build a space shuttle, a lot of basic research on

Fig. 3-2. Researchers used their knowledge of ceramics to create these tiles for the Shuttle. This worker is mounting the tiles on the Shuttle's outer surface.

ceramic materials had already been done. NASA decided to apply what they knew about ceramics to the shuttle. Studies were made to learn if ceramics could be used as the shuttle's outer layer. They found out how thick the layer had to be and which ceramics would work best. Fig. 3-2.

Problem solving and brainstorming are often used during basic and applied research. Doing this helps researchers come up with new and better ideas.

Fig. 3-1. Scientists often do basic research. They try to discover new information.

IMPACT

We have become part of a larger world because of manufactured devices. TVs, radios, and telephones tell us what is happening in other countries far away. Because of books and computer networks, more information is at our fingertips than at any other time in history.

Market Research

Market research is done to learn what customers want or need. Many questions must be answered before a company is willing to invest money in a product. Several methods are used to learn how customers might respond.

Looking at Supply and Demand

Suppose a company wants to manufacture running shoes. First, it learns how many kinds of running shoes are already on the market. This determines the supply. Next, the company learns how many people still need to buy running shoes. This determines the demand. **Supply and demand** describes the relationship between the number of products available and the number of people who want to buy them.

Supply and demand also influence construction. Suppose a company wishes to construct a new apartment building. Is there a demand for apartments in that area? If not, the building may not be filled. The company could lose money.

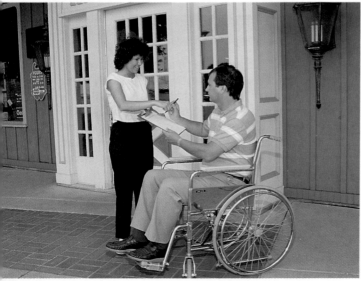

Fig. 3-3. This survey taker is finding out if a consumer likes a proposed product.

If there is not enough demand, usually the product will not be developed. On the other hand, if there is a large demand and a small supply, the manufacturer or builder will be filling a need. The product may then be made.

Taking Surveys

During research and development, many studies may be made of what consumers think and how they behave. A consumer survey may be done. These surveys consist of questions that are asked in person, by phone, or by mail. They are designed to find out how consumers feel about a proposed product. Researchers try to focus on those people who are most likely to buy it. For instance, a survey about ice skates would probably be done in areas that have cold weather.

Studying Behavior Patterns

Researchers also study how consumers behave. For example, they may use sales records from past years to find out what kind of products people bought most. They try to discover patterns. A condominium builder, for instance, might want to learn how well other condos have sold in that town. From this information, predictions are made about what people might buy in the future.

▶▶▶ FOR DISCUSSION ◀◀◀

1. If you were a scientist, would you be more interested in basic research or applied research? Why?

2. Suppose you are a manufacturer who wants to introduce a new child's toy. When you survey consumers, what kinds of questions will you ask?

Extension Activity

■ Suppose the company you work for wants to develop a new product. The product is a mini-VCR, no bigger than a checkbook. The tapes used in it will be the size of a credit card. The new VCR has no other special features and works the same as larger VCRs. Write a list of at least 5 questions you want to ask consumers. Then survey from 3 to 5 people. As a class, compile the responses you collected. Will consumers be interested in buying the new VCR?

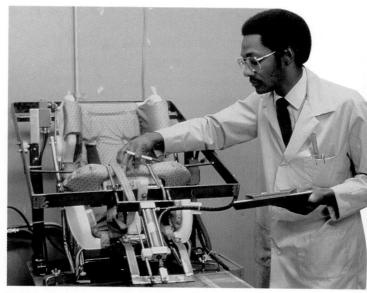

Fig. 3-4. This researcher is trying to improve the way children's car seats are made.

Manufacturing Research

The needs and wants of our society change very fast. In order to be successful, manufacturers must constantly improve existing products. They must also come up with ideas for new products. Fig. 3-3. Many companies set aside a share of their profits for these tasks. Often the investment does not pay off for many years.

The department in a company responsible for new products and methods is called **research and development** (R & D). The department's goal is to satisfy consumers. The product must also satisfy the company's need to make a profit.

The department also looks for ways to improve manufacturing methods. Fig. 3-4. Advanced equipment, such as computers and robots, may be considered.

R & D departments stay informed about new materials. They must consider a material's cost and if it is readily available. Can it be handled with existing equipment at the factory? Sometimes a new material may be so much better that it is worth buying new equipment.

Patents

When a company develops a new product, it usually decides to apply for a patent. A **patent** is the right, granted by the United States government, to produce something. It protects inventors or manufacturers from those who might copy their ideas.

On occasion, different people come up with the same new idea at the same time. R & D departments are responsible for learning if another company has already developed the same product. If there is a patent on the item, no one else may manufacture it unless permission is granted.

▶▶▶ FOR DISCUSSION ◀◀◀

1. Suppose a company makes cotton shirts. If the shirts are sewn with strong thread, they will last at least five years. If they are sewn with weak thread, they will last only two years. The company plans to use weak thread. The shirts will wear out sooner. Then customers will have to buy more. What do you think of this practice? Is it fair to consumers? Under what circumstances might it be fair? When might it be unfair? Would you want to work for this company?

Extension
Activity

■ **List twenty different products that you use every day. Which are made to last a long time? Which are not? Why do you think this is so?**

Developing the Product

After the idea for a product is approved, it must be developed. Designers must put the idea on paper. Models must be made. The product must be tested with consumers.

Design

The idea for a product usually includes what the product will do. What it will do is its function. The idea does not always include what the product will look like. Its form may have to be determined by designers. Form includes its shape, appearance, and the materials from which it is made.

For some products, form is the most important factor. For example, very often the only thing "new" or different about a new car is the way it looks. Many other cars may have the same equipment and features. In this case, its design may determine whether or not people buy it. Fig. 3-5.

For other products, function is more important. For example, how well a laundry detergent cleans is more important than how attractive its package is.

Today, most manufacturers pay attention to both form and function. It is especially important that products be easy for people to use.

Fig. 3-5. The form of this car is an important sales feature. What about its design makes you think of speed?

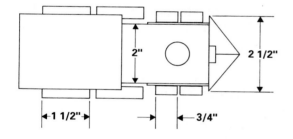

Sketches (Part A)

Fig. 3-6. Sketches (A) are a way of thinking on paper. They're good for trying out ideas. The drawing in part B was done on a computer. This type of drawing is often used in catalogs. Anyone who can read a working drawing (C) could make the item shown. Assembly drawings (D) identify the parts.

Computer-Drawn Pictorial
(Part B)

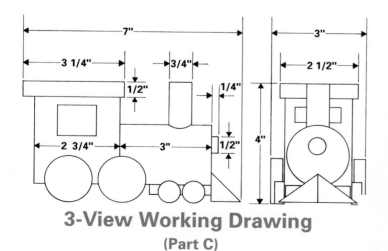

3-View Working Drawing
(Part C)

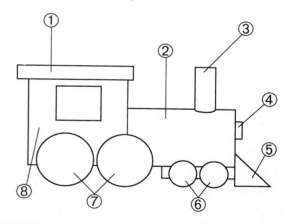

1.	Roof	5.	Cattle Guard
2.	Boiler Area	6.	Small Wheels
3.	Smoke Stack	7.	Large Wheels
4.	Light	8.	Cabin

Assembly Drawing (Part D)

Drawings

To determine a product's design, many drawings are made. Sketches are the first type. They are usually done using pencil and paper. Fig. 3-6, Part A. Engineers review sketches to check if a design can be easily manufactured. From a number of sketches, the best one is selected.

Finished drawings are done next. Sometimes they are made using computers. Fig. 3-6, Part B. The computer can help improve speed and accuracy. Also, any changes in the design can be made quickly and easily. Shades of color and details are shown. One or two final designs are then chosen.

After final decisions are completed, specifications are written down. **Specifications** are information about the product. They tell its size and shape, type of materials used, and any special details. Specifications are placed on working drawings. **Working drawings** show how to make the product. Exact sizes and shapes are shown. The object is usually seen in three ways: from the front, top, and right side. Fig. 3-6, Part C. Assembly drawings may also be created. They show how all the parts are put together. Fig. 3-6, Part D.

Models and Prototypes

Models are often made during the design process. A model is a non-working version of the product. It is usually made to check appearance only. Various materials may be used, such as clay, wood, plastic, or cardboard. The designers want to see how the product will look. Sometimes other characteristics can be studied. From looking at the model, more decisions can be made about the product.

A **prototype** is a full-size working model of the product. It is usually built by hand. Fig. 3-7. The purpose of a prototype is to test the product. The prototype is subjected to much more stress than the actual product would be. For example, a refrigerator may be opened and closed many more times than it would be under normal use. Engineers study the tests. Did the hinges and door seals hold up? Did any other problems develop? Problems can be corrected before manufacturing begins.

Prototypes are often shown to customers. The customers have a chance to give their opinion. This information may result in more design changes.

What Customers Think

Test marketing of a new product may also be done. During a test, samples of the product may be given away, or the product may be sold in only one area of the country. For example, fast-food chains introduce new foods in certain locations. Researchers keep close track of how many people buy an item. This tells the manufacturer if there will be enough demand.

Fig. 3-7. Prototypes are made only of products selected for manufacture. This plane prototype is thoroughly tested before manufacturing begins.

Construction Research

As with manufacturing, important questions must be answered before construction is begun.

Analyzing Needs

Most builders want to know that a structure is really needed. They ask questions to find out. For example, a family may get tired of apartment living and decide to build a new house. Their questions might include the following:

- Do they plan to have more children?
- Is it possible that the parents could receive job transfers?
- Is their current income large enough to pay for the home?
- How will other expenses be affected?
- If they do not already own a car, would they need to buy one?

Businesses also review their needs. A manufacturer may want to relocate to reduce taxes and utility costs. Labor, too, may be cheaper in another area. A retail business may believe that a better location will mean more customers. Fig. 3-8. A business may also want to build a second store.

Communities have to meet the needs of growing and changing populations. Heavy traffic can create a need for new or wider roads. Traffic surveys are done to research how serious the need is. More children in the community may create a need for new or larger schools. Studies can be made to determine which is better, a bigger school or a new one.

Choosing a Site

The **site** is the place on which a structure is built. Many factors must be considered when choosing a site. Fig. 3-9. They include location, zoning, physical characteristics, and cost.

▶▶▶ **FOR DISCUSSION** ◀◀◀

1. **Name some products for which form is more important than function, and vice versa. In each case, tell why you think this is so.**
2. **Can you think of any products which went up or down in price because of supply and demand? Explain what happened.**

Extension **Activity**

■ **As you use different products, examine their design. Select two that seem more difficult to use than they should be. Make drawings showing how they could be improved.**

Company **Activity**

■ **Make a model or prototype of the product your student company plans to manufacture.**

Fig. 3-8. Would this pharmacy attract more customers in a different location?

Fig. 3-9. All these factors are important when choosing a site.

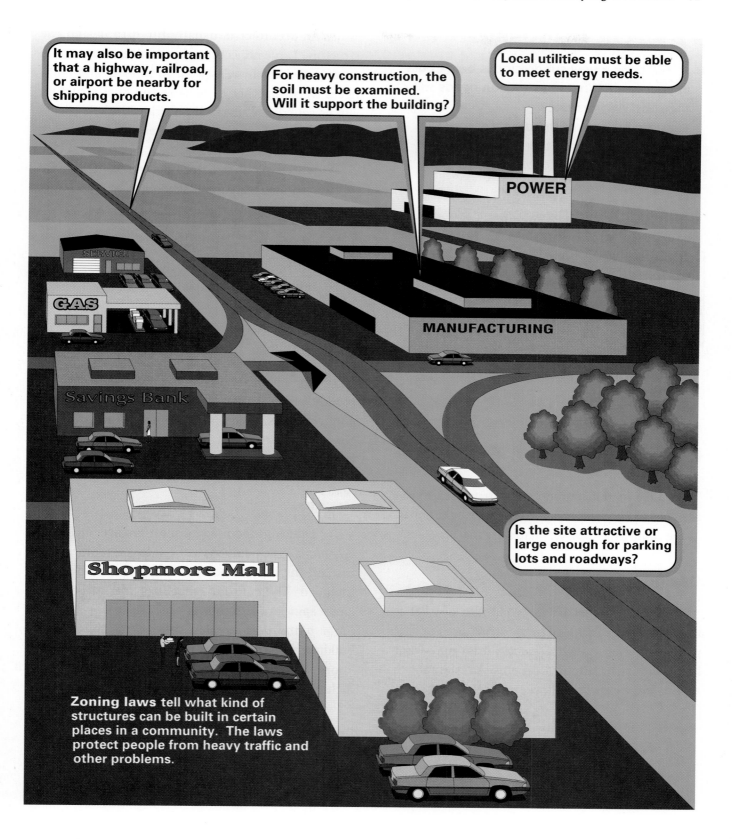

▶▶▶ FOR DISCUSSION ◀◀◀

1. Suppose a new subdivision is being built in your town. What construction needs will the subdivision create?

2. Identify some of the different zones in your community. Are there any areas that seem to be a mixture of zones? Describe those areas.

___

Extension

Activity

■ Take a walk in the area around your school. What construction needs do you see nearby? Find out how a person can make your local government aware of construction needs.

Developing the Project

Like manufactured items, structures must be designed. The plans are put on paper by architects. At the same time, cost estimates are made.

Design

Builders must have a good idea of what features need to be included in the structure. For example, a large family with five children may want a home with at least four bedrooms. A restaurant that expects large crowds will need ample parking space.

An architect is usually hired to design the structure. Large projects, such as bridges, may require an engineering firm as well. Architects and engineers work together to be sure the final structure will be sound. Architects also work with their clients to be sure the structure meets their needs. They ask the client such questions as:

- What type of structure do you want to build — home, restaurant, school?
- Who will use it?
- Why is it needed?
- Where will it be located?
- When must it be completed?
- How much can be spent to build it?

As information is gathered, the architect makes sketches. Many architects use computers to make drawings. As the architect works, the client reviews the sketches and gives feedback.

The client may accept the design as it is or request changes. For large projects, such as shopping centers, a model may be made. This gives the client a better idea of what the finished structure will look like.

Drawings

Drawings required for most projects include:
Site plans. Site plans show the location of the structure on the lot. Boundaries, roads, and utilities are included. The contour (shape) of the lot and any landscaping are also shown.
Floor plans. A floor plan shows how rooms in a building are arranged, as viewed from above. Walls, doors, stairways, and other built-in features are included. Fig. 3-10.

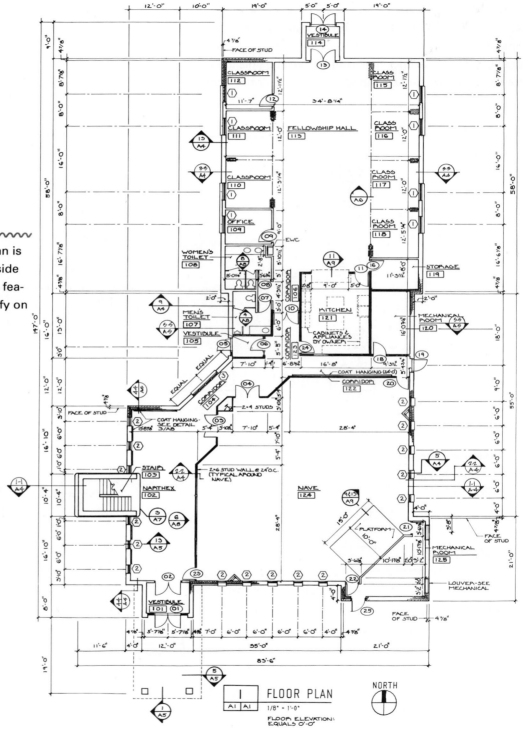

Fig. 3-10. A floor plan is like a map of the inside of a structure. What features can you identify on this drawing?

Fig. 3-11. The building elevation (A) shows features of the exterior. For example, what type of siding will be used on this church? The detail drawing (B) shows how steel columns will be finished with wood framing. The section drawing (C) is of the interior of a column.

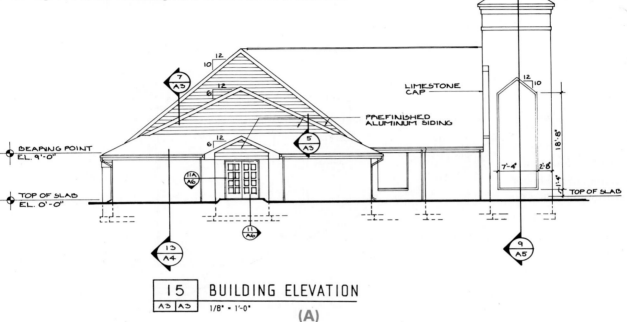

(A)

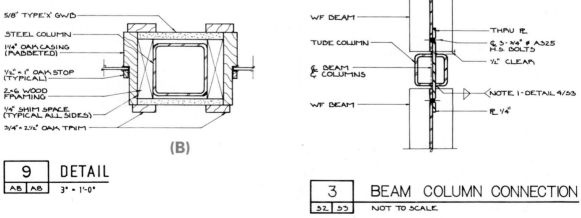

(B)

(C)

Elevations. Elevations show the outside of a structure, as viewed from ground level. A separate elevation is drawn for each side of the structure. For a house, elevations are done for front, sides, and rear. Fig. 3-11, Part A.

Detail drawings. Detail drawings are made of any parts that need to be shown in more detail than was given in other drawings. More information about that particular part is given. Fig. 3-11, Part B.

Section drawings. Section drawings show a slice of the inside of a structure. A wall may need a section drawing that shows such things as floor joists, studs, and rafters. Fig. 3-11, Part C.

System drawings. System drawings show plumbing, electrical, and heating and ventilation systems. Plumbing fixtures and light switches are included in these drawings.

Specifications

Specifications are written instructions about the materials, standards, and governmental regulations that apply to the project. Standards are rules that cover such things as the number of outlets or quality of pipe that must be used in a building. Architects include specifications with the drawings for a structure.

During construction, workers depend on specifications to tell them what materials to use and what rules to follow. Fig. 3-12. Architects refer to the specifications when they check progress on the structure. Building inspectors use them to make sure that standards and regulations are followed.

Fig. 3-12. These specifications list door types, sizes, and framing.

DOOR SCHEDULE

NO.	WIDTH	HEIGHT	TH	MATL	FIN	ELEV	MATL	FIN	ELEV	LABEL	HDWR SET	NOTES
		DOOR					DOOR FRAME					
01	(2) 3'-0"	6'-8"	1 3/4"	WOOD	S & V	B	WOOD	S & V	I	-	1	
02	(2) 3'-0"	6'-8"	1 3/4"	WOOD	S & V	C	WOOD	S & V	II	-	2	
03	(2) 2'-10"	6'-8"	1 3/4"	WOOD	S & V	C	WOOD	S & V	II	-	2	
04	(2) 2'-10"	6'-8"	1 3/4"	WOOD	S & V	C	WOOD	S & V	II	-	2	
05	3'-0"	6'-8"	1 3/4"	IM	PNT	D	WOOD	S & V	III	-	3	IM SIDELIGHT
06	3'-0"	6'-8"	1 3/4"	WOOD	S & V	A	HM	PNT	III	-	8	
07	2'-10"	6'-8"	1 3/4"	WOOD	S & V	A	HM	PNT	IV	-	8	
08	2'-10"	6'-8"	1 3/4"	WOOD	S & V	A	HM	PNT	IV	-	8	
09	3'-0"	6'-8"	1 3/4"	WOOD	S & V	A	HM	PNT	IV	-	6	
10	(2) 3'-0"	6'-8"	1 3/4"	WOOD	S & V	A	HM	PNT	IV	-	5	DOORS TO HAVE 12 x 12 SIDELIGHTS
11	3'-0"	6'-8"	1 3/4"	WOOD	S & V	A	HM	PNT	IV	-	7	
12	3'-0"	6'-8"	1 3/4"	WOOD	S & V	A	HM	PNT	IV	-	6	
13	(2) 3'-0"	6'-8"	1 3/4"	WOOD	S & V	A	HM	PNT	IV	-	5	
14	(2) 3'-0"	6'-8"	1 3/4"	IM	PNT	D	WOOD	S & V	IV	-	4	
16	3'-0"	6'-8"	1 3/4"	WOOD	S & V	A	HM	PNT	IV	-	6	
18	3'-0"	6'-8"	1 3/4"	WOOD	S & V	A	HM	PNT	IV	C	7	
19	3'-0"	6'-8"	1 3/4"	IM	PNT	D	WOOD	S & V	IV	-	3	
20	3'-0"	6'-8"	1 3/4"	WOOD	S & V	C	WOOD	S & V	II	-	9	
21	3'-0"	6'-8"	1 3/4"	WOOD	S & V	A	HM	PNT	IV	C	6	
22	3'-0"	6'-8"	1 3/4"	WOOD	S & V	A	HM	PNT	IV	C	6	
23	3'-0"	6'-8"	1 3/4"	WOOD	S & V	C	WOOD	S & V	II	-	9	
24	3'-0"	6'-8"	1 3/4"	WOOD	S & V	A	HM	PNT	IV	-	7	
25	3'-0"	6'-0"	1 3/4"	WOOD	PNT	A	WOOD	S & V	IV SIM	-	10	
201	3'-0"	6'-8"	1 3/4"	WOOD	PF	A	WOOD	PF	IV	-	6	SEE NOTE BELOW

HM - HOLLOW METAL TH - THICKNESS MTL - MATERIAL FIN - FINISH HDWR - HARDWARE ELEV - ELEVATION
IM - INSULATED METAL S & V - SAND AND VARNISH PNT - PAINT

NOTE:
PREHUNG, PREFINISHED,
HOLLOW, CORE DOOR
AND FRAME.

Estimates

Owners want to know in advance how much a project is going to cost. For this purpose an estimate is made. The **estimate** lists what the approximate costs will be for the entire job. It also must say how long the job will take. People who make estimates are called estimators. They study the drawings and specifications for the structure. Their estimates are based on their own experience, the type of construction, cost of materials, and cost of labor.

One way of making a rough estimate is to find the area of the proposed structure. The area is figured by multiplying the length of the structure times its width. For example, a one-story building that measures 30 ft. by 45 ft. would have an area of 1350 square feet ($30 \times 45 = 1350$). The estimator then looks at the cost of similar structures in the same town. If construction costs are about $65 per square foot, that amount is multiplied times the area. In this case $1350 \times \$65 = \$87,750$.

Rough estimates help owners decide if they can afford to build the structure. If the cost is too high, they may build something smaller or wait until more money is available. These estimates also help architects give advice to owners about what to do.

Bids and Contracts

Finally, the owner must choose a contractor to do the work. A contractor is a person or firm responsible for building the project. Sometimes several firms are asked to bid on the project. Bids are estimates of what the contractor will charge. The bids are based on the drawings, specifications, and owner's estimate for the structure.

For very large structures and public buildings, ads announcing the project are placed in newspapers. Any contractor who wishes to may bid for the job. Often, the bids are kept secret. When all bids have been received, they are reviewed. The work record of each contractor is considered. The job is usually offered to the qualified contractor offering the lowest bid. A contract is signed by both parties that spells out how the job will be done.

▶▶▶ **FOR DISCUSSION** ◀◀◀

1. **Using an example, explain how an architect would use the problem-solving method in designing a structure.**
2. **Why do you think building standards are necessary? What would happen if no standards were enforced?**
3. **What do you think might happen if a large construction project, such as a hospital, ran over the estimated costs?**

Extension Activity

■ **Sketch a floor plan of your home. Include approximate dimensions of every room. Be prepared to discuss the advantages and disadvantages of the layout.**

Company Activity

■ **Do a series of elevations for the structure your student company is planning to build. Include all dimensions.**

Chapter Highlights

● Various kinds of research are used in deciding on products. They include basic, applied, and market research. Market research includes looking at supply and demand, taking surveys, and studying behavior patterns.

● After a product idea is approved, it must be developed. Drawings and models are used.

● During manufacturing research, test marketing may be used to see if customers like the new product. Patents are also reviewed.

● Construction research involves analyzing needs and choosing a site. Factors in choosing a site include location, zoning, physical characteristics, and cost.

● A construction project is developed using drawings and specifications. Then estimates are made.

● Contractors submit bids on construction projects. The lowest qualified bidder is usually given the job. A schedule for the project is then prepared.

Test Your Knowledge

1. How do basic and applied research differ?

2. What methods are used in market research?

3. Which department in a company looks for new products and ways of making things?

4. What is a patent?

5. What are specifications and where are they found?

6. What is a prototype and why is it used?

7. List the four factors in choosing a building site.

8. What are the different building zones a community may have?

9. How do rough estimates help owners?

10. Describe how a contractor is selected for a large project.

Correlations

LANGUAGE ARTS

1. Go to the reference section of your school library. Using a recent edition of an almanac, find the number of patents given to inventors in 1990 and 1991 by the United States Patent Office.

2. Examine a product advertisement. Does the ad stress form or function? Make a list of nouns, adjectives, verbs, and adverbs used in the ad.

SOCIAL STUDIES

1. Write a report on the history of the patent process. Include the story of at least one inventor.

SCIENCE

1. Some materials used to make products are chosen for their strength. Make a strength test of three different sewing threads. Tie a length of thread to a plastic bucket. Weigh the empty bucket. Gently pour in a quantity of sand. After the thread breaks, weigh the bucket and sand. Repeat the experiment with each kind of thread. Which held the most weight before breaking?

MATH

1. Suppose you built a model car with a scale of 1:7 (1 foot = 7 feet). The model you build is 2' long. How long would the actual car be?

You·Can·

— Make a Difference —

Student Inventions Answer Many Needs

Do you like to look for new ways to solve old problems? Do you like to invent things? For a lot of young people, the answer to those questions is "Yes!" Many of them have taken part in *Invent America!*, an invention "contest" for students from kindergarten through eighth grade. Through *Invent America!*, students have created inventions that save time, money, the environment, and even lives. Some of the winners have become real entrepreneurs and are selling their inventions.

At the state level, one student is chosen per grade to go to the regional competition. Those winners then compete at the national level. The

national winners' inventions are placed on display at the Smithsonian Institution in Washington, D.C. Winners have also represented the United States in international competition. More than 30,000 schools have taken part in the program.

The inventions cover many areas and fill a wide variety of needs. Here are a few.

When in the fifth grade in Evergreen, Colorado, Casey Golden was a 1989 national winner with his biodegradable golf tee. His tee, which blends into the ground naturally, not only saves trees that would be cut to make wooden tees, but also helps the environment. Casey has sold tens of thousands of his tees already.

A 1990 eighth-grade winner, Larry Villella of Fargo, North Dakota, has sold more than 2,000 of his Conserve Sprinklers. These C-shaped sprinklers fit around trees and shrubs to water the roots. They have a built-in water regulator as well.

Brad Bolerjack, a 1989 sixth-grade winner from Lockport, Illinois, read in the newspaper that many people got hurt trying to jump start their cars. As a result, Brad developed a set of safe and efficient jumper cables.

Katherine Szudy, a 1990 eighth-grade winner from Parma, Ohio, got the idea for her invention one day when she saw a blind man on television trip and fall down a flight of stairs. Her invention, called the Braille Rail, puts braille symbols on stairway hand rails. Blind people can read the symbols and find out if they're about to climb up or down. They're also told which step they're on and how many steps are ahead.

Seventh-grader, Jennifer Shelburne of Bardstown, Kentucky, won a 1991 Environmental Award for her Cardboard Filler Machine. She learned that the factory where her mother works baled cardboard for recycling but spent $13,000 a year for packing materials. She invented a machine that shredded the cardboard and created packaging filler that is safer for the environment.

A 1991 seventh-grade winner, Reed Wangerud of Littleton, Colorado, invented the Treetop Tracker to help lost hikers. The invention uses the principle of model rocketry to fire streamers into the treetops. Searchers looking for the hikers from airplanes can see the streamers more easily than they can see a person on the ground.

To help bridges withstand earthquakes, 1991 eighth-grade winner William Cotten of Red Oak, Texas, developed the Seismically Activated Fracture Eliminator (S.A.F.E.). His invention uses a combination of ballbearings and stretchable bands.

Invent America! was started in 1987 by then Vice President George Bush. It is sponsored by the nonprofit U.S. Patent Model Foundation. With the help of large corporations, *Invent America!* provides free educational materials, teacher training, and more than $600,000 each year in awards for students, teachers, and schools.

WHAT PRODUCTION SYSTEMS NEED

Activity Brief
Designing a Home on the Moon
PART 1 : Here's the Situation...........

The chapters in this section discuss some of the things production systems need in order to operate. For instance, you will learn about many of the materials used to make products. You will also find out about the tools and machines used to process those materials. While you are learning, you will be solving your own production problem involving materials.

Imagine that you are a member of the first colony of earth people to live on the moon. Engineers have already constructed a large dome that will cover the entire community. Beneath the dome there will be air to breathe. A filter will screen out the sun's most harmful rays during the moon's two-week-long "days." Water will also be supplied, but in limited quantities. All colonists have been asked to submit plans for permanent homes. The best plan will be chosen to build the community.

PART 2: Your Challenge..........

For this activity, you will design a home for moon colonists. What do you have to work with? Unfortunately, not much. Nothing grows on the moon, so there is no timber available. It is also impossible to ship traditional building materials from earth, unless they are extremely light and compact. To succeed with your design, you will have to be creative.

PART 3: Specifications and Limits..........

Your moon home will need to meet certain standards. Read the following specifications and limits before you begin.

1. Your structure must be made of materials that actually exist or that can be made from what is found on the moon. In other words, you can't "invent" a make-believe material to do the job.

Also, the materials cannot violate the conditions set forth in the activity. For example, bricks cannot be shipped from earth. They would be too heavy. (See table, "Conditions Under the Dome.")

2. Your home must:
    *   accommodate a family of four comfortably
    *   include areas for sleeping, living, dining/cooking, and bathroom and storage facilities; all areas may be multi-purpose
    *   make provision for conditions on the moon, including climate, light, lack of water, reduced gravity, etc.

3. You must hand in the following:
    *   a daily log sheet of your work
    *   a scale drawing of the exterior of the structure (front, back, and sides), including dimensions
    *   a floor plan of the interior, giving dimensions of rooms
    *   a list of the materials and machines required to create the home

Conditions Under the Dome

Temperature: **95° during the day; 0° at night (except in the greenhouse)**

Water allotment: **7 gallons per person per day; 3 quarts of this is drinking water and cannot be used for any other purpose.**

Growing in the greenhouse: **a rain forest that includes kapok, white pine, bamboo, ferns, fast-growing hardwoods, philodendron, and the section devoted to cactus; dead leaves are recycled as compost; seeds and cuttings may be taken by colonists.**

Tools and machinery: **excavators (backhoe, trencher, bulldozer), crawler crane; all hand tools described on pp. 114 - 115 ; power tools and machines described on pp. 118 - 120 ; machines have been adjusted to work under moon gravity.**

Shipments from earth: **one package weighing no more than 100 pounds and taking no more than 3 cubic feet of space is allowed per person.**

- a one- or two-page report describing the structure, explaining how the materials, tools, and machines will be used during construction, and giving reasons why you chose them
- a list of any resources you used (books, magazines, museum displays, interviews with people, etc.)

4. You will be evaluated on the basis of:
   - originality in the use of materials, tools, and machines
   - how well you followed design specifications
   - how well you took conditions on the moon into account
   - how well you researched your design

PART 4: Materials..........

There are many materials you might use to create your design. Here is a list of possibilities.

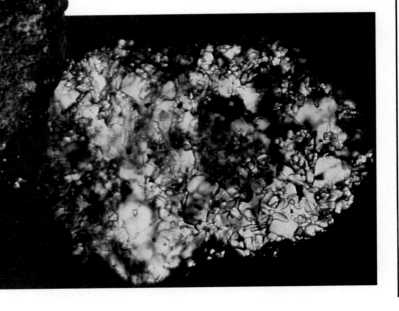

Materials

pencils and paper
drafting tools and equipment, such as T-square, triangles, etc.
computer and CAD software

These photos are of actual rocks brought back from the moon by astronauts. They resemble rocks from volcanoes on earth. The holes were probably caused by gases escaping. Much lunar soil contains tiny colored glass balls.

Safety Notes

- What kinds of safety problems might occur during construction on the moon? You might want to include safety items in your list of equipment.

PART **5**: Procedures...........

The design you choose will be up to you. Still, there are certain steps you can follow that will make your work easier.

1. Be sure to keep a daily log of your work.

2. Read all the information provided with this activity. Be sure you understand conditions on the moon and under the dome. Review the steps in problem solving and use them as you work.

3. Using your school or community library, do some research for your design. For example, you might look up the history of architecture. Designs or materials used in times past may give you some good ideas.

4. Make some sketches of possible designs. Try to take all moon conditions into account. Make lists of materials and equipment needed for each design.

5. Select the best design and make a finished drawing.

6. Make a floor plan.

7. Write your tools, machines, and materials list, your report, and your list of resources.

PART 6: For Additional Help...........

Information about the following is found on these pages in the text:

You might also want to research these related subjects in reference books:

Consider also magazines, such as *The Futurist* and *Popular Science*.

Research Tip: Investigate housing in desert climates.

PART 7: How Well Did You Meet the Challenge?............

When you've finished designing your structure, evaluate it. Does it do what it's supposed to do? Ask yourself the following questions. Your teacher and classmates may take part in this evaluation.

1. How well does your structure solve the problems of construction on the moon?

2. How well would your structure hold up under conditions there?

3. Is your drawing clear and understandable? Are the correct views shown? What about your report? Is it clear and easy to understand?

4. Did you do enough research on your subject?

PART 8: Extending Your Experience...........

Think about the following questions and discuss them in class.

1. What conclusions can you draw from this exercise about materials, tools, and building processes?

2. Would your shelter design lend itself to mass production? How could mass production be handled on the moon?

3. How would a sanitation system work in a community based on your design? A communication system?

4. What air pollution problems might exist under a dome?

5. What sort of streets and sidewalks would a moon colony have?

6. Who in real life is working on designs for living on the moon?

Conditions on the Moon

Distance from earth:
221,456 miles
Diameter:
2,160 miles
Surface area:
14,650,000 square miles

Rotation period:
27 days, 7 hours, 43 minutes
Length of day and night:
14 days each
Temperature at equator:
260° when sun is at zenith;
280° below zero at night

Surface gravity:
1/6 of the earth's gravity
Atmosphere:
little or none; no clouds, no wind
Soil content:
chunks and ground-up rock
containing aluminum, iron, silicon,
basalt, and tiny balls of glass

CHAPTER 4

Information, Energy, and Time

Introduction.....................................

As you have already learned, every production system needs input from these seven basic resources: people, capital (money), information, energy, time, materials, tools and machines.

Each input to a production system depends on the other inputs. For instance, without materials, products could not be made. Without information about these materials and the methods used to process them, production systems could not work well.

In Chapter 2 you learned about capital and people. In this chapter you will learn more about information, energy, and time. Chapter 5 will discuss materials, and Chapter 6 will cover tools and machines.

After studying this chapter, you should be able to.....................

Discuss several kinds of information required by production systems.

Tell where energy comes from and name the six forms of energy.

Explain the differences among limited, renewable, and unlimited energy sources and give examples of each.

Tell how energy is used in technological systems and how it can be conserved.

Describe time and motion studies and discuss their importance to production systems.

Words you will need ....

bioconversion
cogeneration
energy
Environmental
　Protection Agency
flow chart
fossil fuels
kinetic energy

Occupational Safety
　and Health
　　Administration (OSHA)
photovoltaic cell
potential energy
schedules
time and motion
　studies
work

Information

Can you answer this riddle? *You can give it away and still keep it. It weighs nothing itself but is often stored in containers that weigh hundreds of pounds. It can be sent thousands of miles on a beam of light.* What is it?

Did you guess "information"? If you did, you're right. Information is an important resource of production systems. Without it, they could not operate. Production systems use many kinds of information.

Product History

Product history is information about products made in the past. The information is then used in making new designs. Completely new products are rare. Most are based on other, similar products.

For example, the design of the space shuttle is based on that of airplanes. Fig. 4-1 (pgs. 78-79). The first airplanes were gliders. These planes used wind to make them fly. In 1903, the Wright brothers designed and flew the first engine-powered aircraft at Kitty Hawk in North Carolina. As new planes were developed, the information about plane design grew. Seventy years after Kitty Hawk, engineers used this information to create the first space shuttle. The product history of the shuttle goes back a long way.

Market Research

As you learned in Chapter 3, production systems learn about customers' wants and needs through market research. Telephone surveys are one research technique. People are called at home and asked questions about different products. This helps determine their buying habits. Also, a questionnaire may be mailed to customers who recently bought certain products. The information gathered helps a manufacturer decide what changes to make in future models of the product.

Construction companies also need information about the marketplace. Suppose a company wants to buy a certain piece of land. What are the zoning laws? Fig. 4-2. What are housing prices like nearby? What kinds of housing are needed? The answers to these questions will determine whether the company could make a profit. After all the information is analyzed, a decision is made.

Construction Site Data

Contractors need to gather information about proposed construction sites. They may want to learn the best area in which to build. In the case of roads pictures may be taken of the site from the air. Some sites may be more difficult to use than others. They may be too hilly or sandy, for example. This would increase costs. Another site may have to be chosen.

Fig. 4-2. What kind of structures may be built on this site?

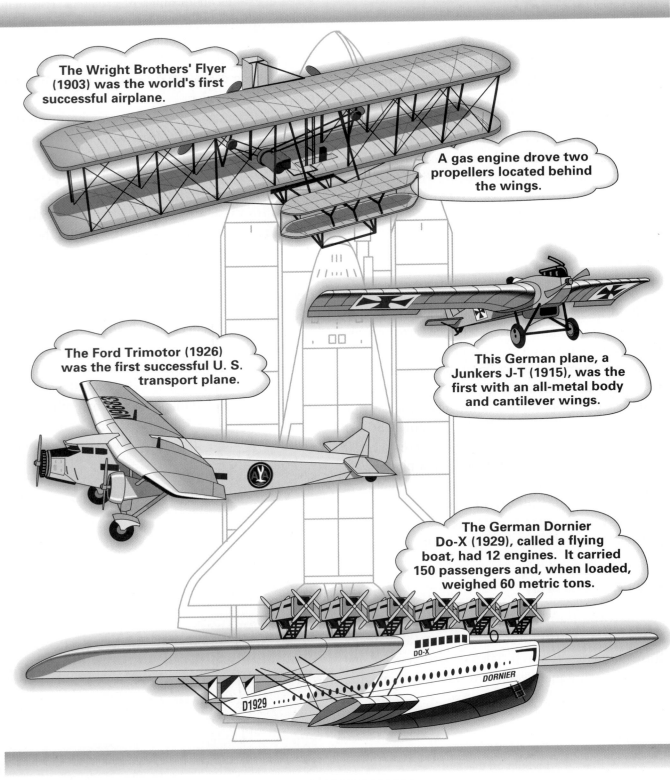

Fig. 4-1. The design of the space shuttle grew out of designs for airplanes. What similarities can you find?

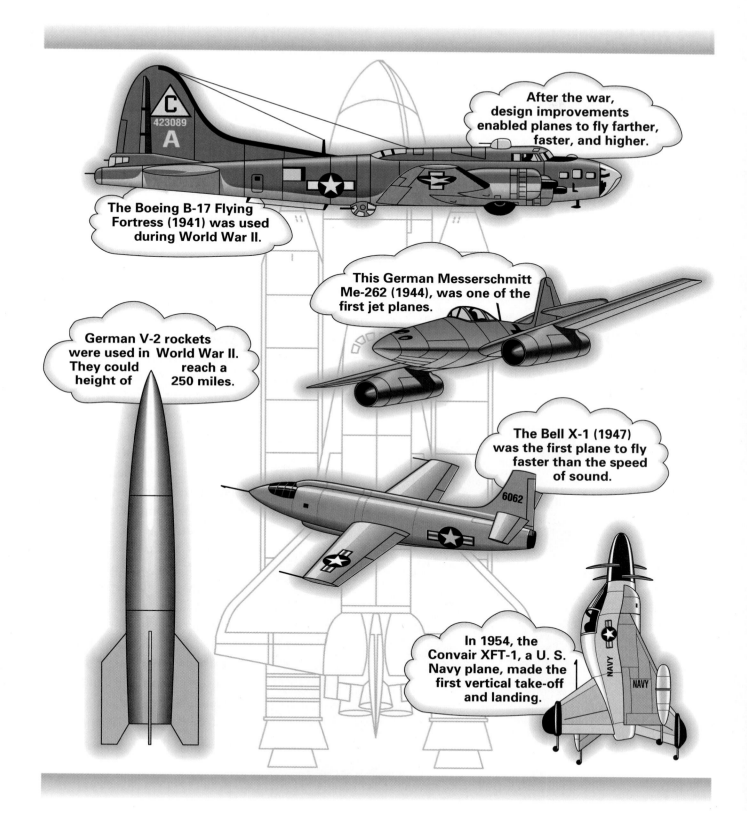

Laws and Regulations

A number of laws regulate production systems. Some have to do with the age of workers. A long time ago young children were put to work in "sweatshops." There they worked long hours under unhealthy conditions. This is now against the law.

Another law sets the minimum hourly wage for workers. An employer cannot pay a worker a salary that is lower than the minimum wage.

The **Occupational Safety and Health Administration** (OSHA) was established in 1970. Its regulations protect the safety of all workers. Most companies follow OSHA rules carefully. If an OSHA inspector finds a violation, however, the company may be fined.

In recent years a number of laws have been passed that protect consumers. Production companies must design products so that they are safe to use. For example, automobile bumpers must resist a certain amount of impact. Toys must be safe for children to play with.

Zoning laws are important for construction companies. The laws protect a community from structures that may not be right for an area. For example, a store could not be built in an area zoned for homes.

Building codes affect the quality of a structure. The codes tell what parts are needed and where they should be placed. Electrical wiring and plumbing are both covered by building codes. Fig. 4-3. During construction, an inspector visits the site to make sure the structure is being built correctly.

Many laws have been passed that protect our natural environment and resources. The **Environmental Protection Agency** (EPA) enforces the laws and identifies dangers to the environment. A number of these laws affect production systems. For example:

- Manufacturers must reduce the harmful pollutants coming from smokestacks.
- Factories must reduce, eliminate, or process their harmful wastes. Dangerous chemicals may not be dumped into waterways or landfills.
- Products themselves may be regulated. For example, a car's engine may emit only a certain level of pollutants.
- Construction companies must file an Environmental Impact Statement. This tells how the proposed project will affect the surrounding area. If the effects could be harmful, the project may not be allowed.

Fig. 4-3. The wiring this electrician is doing must follow the building code.

News of Improvements

New materials are always being developed. New tools and methods are created to process the new materials. Production companies must stay informed about what's new. They do a lot of research. Then they use the information to improve their operation. This helps them keep up with other companies. They may find ways to use their money more efficiently. They have a better chance of making a profit.

For example, in the past the auto industry used many workers to weld, bolt, and paint cars. Today, much of that work is done by robots and other machines. The robots have reduced labor costs for car makers. The money that was spent on salaries, health benefits, and pensions is now used to set up robot production lines.

Production Schedules

Production **schedules** indicate when different stages of production should begin. They also tell when certain materials are needed. To make a schedule, a production manager gathers information from many sources. Meetings are held with designers and production workers. Suppliers indicate when materials will be delivered.

Time and motion studies are also sometimes used to plan schedules. These studies measure the time and movements needed to complete each operation in the production schedule.

All this information is used to develop a flow chart. A **flow chart** identifies all the operations needed to make a product. Fig. 4-4. Symbols used to represent different steps make the chart easier to read. The steps are given in the correct order. The reader can tell how the work will flow. If delays occur during production, the manager makes changes in the flow chart and schedule.

Feedback

As you learned from the universal systems model, feedback is information about system output. Feedback can be used to improve the system.

For instance, a lot of stores may call the manufacturer saying they are sold out of a product. They want to order more. The production team may increase production based on this feedback. Other feedback may come from customers who indicate that a product is defective. The production team will have to find the problem and fix it.

Fig. 4-4. If you follow this chart, you know in what order steps must be taken during production.

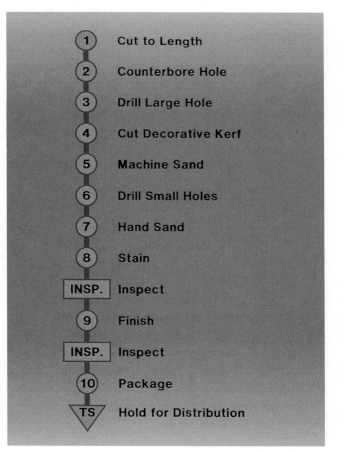

▶▶▶ FOR DISCUSSION ◀◀◀

1. Imagine that you have been asked to take part in a market survey for a new brand of cereal. What kinds of questions do you think you will be asked? What if the survey were for running shoes? How would the questions differ?

2. Children are no longer allowed to work in factories. Do you think this is a good idea or not? Why?

3. Is your class manufacturing a product? What steps will need to be shown on your schedule and flow chart?

Company Activity

■ If your class is manufacturing a product, use resources to research the history of the product. Describe how it has changed or improved over the years. Has it been replaced by new products?

Company Activity

■ Using a working model of your class product, collect market information. Ask potential customers how they feel about the product. Would they buy it? If not, why not? What changes would they suggest?

Energy

Production systems use energy for lights and heat, transportation, processing materials, and operating equipment.

Energy is the ability to do work. When you have plenty of energy, you can get a lot of work done. Scientists define **work** as moving a force over a distance. You have probably heard the term "horsepower" applied to car engines. Horsepower (hp) is the unit used to measure power supplied by a machine. Horsepower tells how much work the machine can do.

All the energy on earth came originally from the sun. The sun is a natural source of nuclear energy. The heat and light given off by the sun are tremendous.

Both plants and animals need the heat and light the sun provides. Plants use that energy to grow and produce flowers, seeds, leaves, and fruit. When we eat plant foods, we are really receiving energy from the sun. After plants die, they are changed over millions of years into coal, oil, and natural gas.

Energy sources are all around us. There are, basically, three kinds: unlimited, renewable, and limited.

IMPACT

The need to conserve energy has brought about new trends in construction. They include fewer and smaller windows, solar-assisted heating, and better insulating materials.

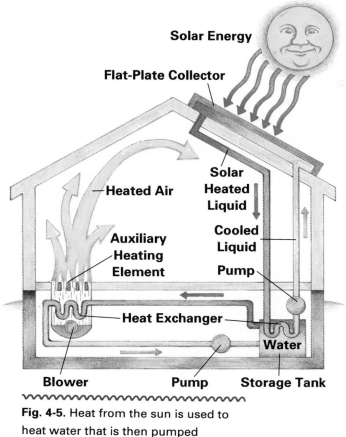

Solar Energy

Flat-Plate Collector

Heated Air

Solar Heated Liquid

Auxiliary Heating Element

Cooled Liquid

Pump

Heat Exchanger

Water

Blower   **Pump**   **Storage Tank**

Fig. 4-5. Heat from the sun is used to heat water that is then pumped throughout the house.

Unlimited Sources of Energy

Some energy sources will last as long as the earth lasts. They are unlimited, and it does not matter how much of them we use.

Sun

Technology can make direct use of the sun's light and heat. **Photovoltaic cells** (solar cells) change the energy of light into electricity. This is useful where regular electricity is not available, such as for satellites.

In the California desert, a solar power system has been installed. Tall, curved mirrors focus the sun's rays on a black pipe filled with liquid. The sun heats the liquid as high as 725° F. This produces the steam needed to make electricity in a power plant.

Solar energy can also be used for heating buildings. **Passive solar buildings** are heated when energy is absorbed through windows. **Active solar buildings** have solar collectors. The collectors heat water. The water is used in heating units and for cooking and bathing. Fig. 4-5.

Wind

In years past, windmills were used to pump water and mill grain. However, wind could not be depended upon. Other energy forms became more popular. Today wind energy is once again important. Groups of generators called wind "farms" are being used to produce electricity in California. Fig. 4-6.

Water

Water is another plentiful source of energy. Most early factories were located near rivers. The moving water turned huge waterwheels. Attached to belts and pulleys, the wheels supplied power to machines. Today, moving water flowing over dams through generators produces hydroelectric power. Hydroelectric power provides from 10 to 15 percent of the electricity used in the United States.

Fig. 4-6. Wind farms, such as this one at Altamont Pass, California, can generate enough electricity for 150,000 people.

Earth Heat

Energy that comes from inside the earth is called geothermal energy. In places where volcanoes are common, this heat is close enough to the surface to use. In many cases, springs under the ground become hot. The springs create geysers. Fig. 4-7. A **geyser** is a column of hot water that shoots up from the earth. North of San Francisco geysers are used for 17 power plants. These plants provide electricity for over a million people.

Renewable Sources of Energy

If certain energy sources are managed carefully, they can be renewed. Wood is an example. However, cutting too many trees can destroy our source of wood. Fig. 4-8.

Trees and Other Plants

For centuries, wood was the major source of energy for heating and cooking. Later, other fuels became more popular. Today, however, wood is again being used in some places. Unfortunately, it can cause high levels of air pollution. Also, trees are renewed very slowly, and forests require a lot of space in which to grow.

Other plants, such as sweet sorghum, sugar beets, and grain can be used to help create gasohol. Gasohol is a mixture of gasoline and plant alcohol. It can be used in place of gasoline. It causes less pollution. The plant sources for gasohol are easily renewed. They grow quickly.

Bioconversion

Energy can be produced by burning animal and plant wastes. The waste used is called biomass. The process is called **bioconversion.** Food product waste, animal wastes, paper, cardboard, and wood can all be used for bioconversion.

Humans and Animals

Whenever you use a hammer to pound a nail or a saw to cut wood, you are using human energy. In some countries, where there are few machines to do the work, human energy is important to production.

Fig. 4-7. Heat radiating upwards raises the temperature of underground rock and water. The heated water or steam expands and rises to form a geyser.

HOW GEYSERS ERUPT

Geyser Basin

B. Water seeps upward.

A. Water at the bottom is far above the surface boiling point of 212 F. (100 C).

C. Boiling water expands into steam and causes an explosion. The geyser spouts.

OLD FAITHFUL GEYSER

For centuries, animals were also used for work. Horses and oxen were often hitched to plows or wagons. Today in some places, these animals work side by side with modern tractors.

Limited Sources of Energy

Certain energy sources are limited. Once they are used up, they cannot be replaced. Almost all the energy we use today comes from such sources.

Fossil Fuels

Fossils are the remains of plants or animals that lived millions of years ago. Some of this plant and animal life was buried in swamps. Over time, high pressure and temperatures turned it into coal, oil, and natural gas. Coal, oil, and natural gas are referred to as **fossil fuels.** Fig. 4-9.

Coal. In the early 1700s, coal replaced wood as a source of fuel for industry. Coal was more efficient.

Fig. 4-8. Renewable sources of energy include crops, such as grain, human and animal muscle power, and wastes that can be burned.

Fig. 4-9. Fossil fuels come from plants and animals that were buried ages ago in swamps and then placed under high pressure and temperatures.

1. As plants and animals died, they formed a layer of organic matter. This layer was covered by more sediment.

2. More plants and animals died, forming layers. High pressure and temperature changes produced peat.

3. More pressure, high temperatures, and shifts in the earth changed the peat into coal, oil, and natural gas.

4. Coal, oil, and gas mining tap these important resources.

Some coal is still used in steel production. Today, however, it is mostly used for generating electricity. Unfortunately, coal is a major source of air pollution.

Oil. Most of the energy used for transportation systems comes from oil, or petroleum. The automobile and petroleum industries grew together. Millions of homes are now heated with petroleum products. It is also an important part of such products as plastics and fertilizers.

Natural Gas. Natural gas is commonly found along with oil deposits. It is a clean fuel and easy to transport. Studies show that there are more natural gas reserves in the United States than formerly believed. The use of natural gas as a fuel is expected to increase. Of all the natural gas we pump for heat and power, industry uses one-third.

Uranium

Uranium is a substance used to produce nuclear fission, a form of nuclear energy. Atoms of uranium are split apart in nuclear reactors. Huge amounts of energy are released. The United States' supply of one type of uranium is limited. A new reactor that uses a different type of uranium is being developed. This second type of uranium is plentiful.

Not everyone approves of the use of nuclear energy. Nuclear reactors are expensive to build. They create dangerous wastes. It is difficult to find sites for safe storage of these wastes. Also, reactors themselves can be dangerous. An accident occurred at the Chernobyl reactor in the Soviet Union in 1986. Radioactivity escaped into the surrounding area. Many people have become ill as a result. The effects of this accident will not be fully known for many years.

Forms of Energy

There are six forms of energy. They are all related to one another. Most forms can be changed into most of the other forms.

Mechanical. Moving objects have mechanical energy. It is the energy of motion. A hammer driving a nail and a book falling from a table have mechanical energy.

Heat (thermal). Heat energy is also from motion — the motion of moving molecules. The more there are and the faster they move, the greater the heat energy.

Light. Without light energy no plants or animals would grow. As you have learned, photovoltaic cells can convert light energy into electricity.

Chemical. Chemical energy is produced by chemical changes in a substance. A battery contains stored chemical energy. When the battery is used, the chemical energy is changed to electrical energy.

Electrical. Electrical energy is the flow of electrons. Electrons are tiny particles contained in atoms. We constantly change electrical energy into other forms.

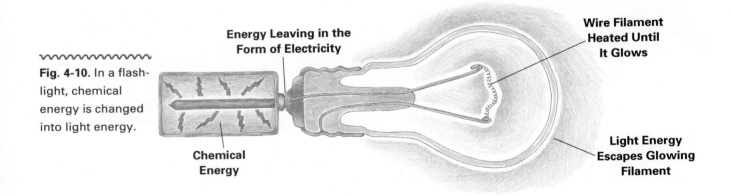

Energy Leaving in the Form of Electricity

Fig. 4-10. In a flashlight, chemical energy is changed into light energy.

Chemical Energy

Wire Filament Heated Until It Glows

Light Energy Escapes Glowing Filament

Fig. 4-11. The force of moving water turns the blades of these turbines in this hydroelectric plant in Oregon. The turbines then generate electricity.

We change it into light with lamps, heat with furnaces, and mechanical energy with motors.

Nuclear. Nuclear energy occurs when matter is changed directly into energy. Only a very small amount of matter is used to produce a lot of energy. The most common use for nuclear energy is in power plants.

Energy Can Be Changed

Often, energy must be changed from one form to another before it can be used. For example, chemical energy in a battery is changed into electricity when the battery is put to use. In a flashlight, the energy is changed once again — into light. Fig. 4-10.

Energy that is stored is called **potential energy**. The energy stored in wood logs is potential energy. Energy in motion is called **kinetic energy**. The heat flowing out of burning wood is kinetic energy. Potential energy must be changed to kinetic energy before we can use it.

How Production Systems Use Energy

The most versatile form of energy available is electricity. About 30 percent of the total energy used in the United States goes to produce electricity. Production systems use electricity for lighting, heating, cooling, cooking, and operating equipment.

Most electricity is generated by turbine generators. A turbine is an engine that rotates when a force is applied. Hydroelectric and steam-electric methods are used to drive the turbines. In hydroelectric plants, water stored behind a dam is released and turns the turbine blades. Fig. 4-11. In steam generating systems, water is heated with fossil fuels or by nuclear fission. The resulting steam turns the turbine blades.

When a single energy source is used for two purposes, it is called **cogeneration**. For example, the steam used to drive a turbine in a power plant may also be used to heat buildings. Waste burned at paper and lumber mills can produce the steam and electricity used to operate equipment. The use of cogeneration is expected to increase sharply in the future.

Fig. 4-12. This steel had to be heated to 2500° F in order to melt it.

Manufacturing

About 40 percent of our country's energy is used by manufacturing industries. Energy costs are important to industry. Sometimes manufacturers relocate to areas where the costs are lower.

Companies that change raw materials into industrial products, such as steel, aluminum, glass, and petroleum products, use large amounts of energy. Other needs for energy include:

- Heat for welding and soldering
- Electrical, fluid, or air power for grinding, drilling, sawing, and shaping
- Heat for furnaces that condition metals, fire ceramics, and dry lumber
- Heat for melting plastics and metals before they are poured or forced into molds. Fig. 4-12
- Power to operate presses that form metal

Construction

Companies that make construction materials, such as concrete and roofing shingles, require energy for production. At construction sites, electricity provides heat, light, and power for machines and tools. Fig. 4-13. Gas and diesel fuel supply power for trucks, cranes, and earthmoving equipment.

Transportation

About 25 percent of our country's energy goes into moving people and goods. Conveyors, guided robot vehicles, and lift trucks move items within factories. Industrial materials are shipped by truck or rail to manufacturers. Entire buildings or parts of buildings may be brought to construction sites using trucks.

Agriculture

Efficient agricultural systems have helped increase our standard of living. Less than three percent of our population are farmers. Yet they produce more food than we can use.

Tractors and other farm equipment depend on fuel for power. Farm buildings require heat and light. Farm products are shipped to food processors in trucks that require fuel. After processing, the food is transported to stores and restaurants.

Conserving Energy

If we want our energy supplies to last, we must use them wisely. Those that are unlimited or renewable must be used in place of those that are limited. Damage to the environment from some energy sources must be reduced. New methods of production must be found to meet our needs for electricity.

Energy conservation can be practiced by everyone. We can add insulation to our homes and replace drafty windows to reduce heat loss. We can seal cracks and replace inefficient heating and cooling systems. Industry can take the same steps. Heat and light can be controlled using computers. More efficient equipment can be purchased. The money saved in energy costs will help pay for these new machines.

Fig. 4-13. How many uses for energy are shown in this picture?

▶▶▶ FOR DISCUSSION ◀◀◀

1. Describe the sun's role in energy.
2. How do you feel about the use of nuclear fission as a source of energy? Do you think it is worth the risks? Give your reasons.
3. Imagine you are going to build a swimming pool in your backyard. What kinds of energy would be used to complete the project?

Extension Activity

■ Find out how the electricity supplied to your home is generated. Draw a system diagram to describe it. Include input, process, and output. List several possible resources for each of the inputs. Explain the role of feedback.

Time

Time is a measured period during which something happens. Time has value. Some workers are paid by the hour. The amount of time they work determines their pay. Other workers may be paid a certain amount for completing a job. The longer the job takes, the less value their time will have.

Time may determine how a project turns out. What will happen if you walk across a freshly painted floor that has not had time to dry?

Everything we do takes time. How much time it takes depends on the materials, processes, and people involved. Fig. 4-14.

Time and Production Systems

Production systems need time to make a product. The amount of time required affects the cost of the product. As you already know, schedules are an important source of information during production. Schedules are a way of controlling time.

The time allowed must match the production system's ability to make products. If the process is too fast, more defects in products may occur. Correcting the defects will take time. If the process is too slow, the cost for making the products will go up.

Many new technologies reduce the amount of time required to perform an operation. For example, computers speed up some tasks. Automated machines run by computers make manufacturing operations faster. Computer-aided design (CAD) systems allow engineers and architects to draw on a computer screen. Changes in drawings can be made more quickly.

For construction, computers analyze stresses and loads affecting structures. Computers can also analyze the shape of the land used for roads. Without computers, all these jobs would take much longer.

Managing Time

In the late 1800s, Frederick W. Taylor, a steel company employee, devised ways to improve manufacturing using scientific principles. One of these ways was a time study. In a time study the time it takes to do a certain task is measured. Today, Taylor is known as the father of industrial management. Fig. 4-15.

Fig. 4-14. By showing the worker how to do the job faster, this supervisor is actually helping him earn the maximum amount per hour.

Fig. 4-15. This computer is using seismic data to help identify drilling locations for an oil company. The computer does the analysis much faster than human workers can.

People who work in time management are called efficiency experts or efficiency engineers. They figure the amount of time needed for each step in the production process. In order to save money, they try to reduce the amount of time a step will take.

Two other pioneers in industrial engineering, Frank and Lillian Gilbreth, studied motion. A motion study is concerned with a worker's movements. The Gilbreths tried to discover the best sequence (order) of movements in doing a job. As a result, standardized movements were established for all workers doing similar tasks. The Gilbreths combined their motion studies with Taylor's time studies.

Time and motion studies have been used by production systems ever since. To make such a study, a trial run or computer simulation may be used.

Industrial robots are programmed based on time and motion studies. The computer moves the robot in a sequence that takes the least amount of time. The faster the robots are, the faster production will be.

Computers have helped manage time in other ways. Records and schedules are maintained on computers. Data can be called up instantly. Changes can be easily made.

Manufacturing

In manufacturing, the fewer the operations, the less time required to make a product. Products are developed by teams of designers and engineers. The designers decide on the use and appearance of the item. Manufacturing engineers select the processes and sequence used to make it. The main objective of these engineers is to limit the number of operations needed.

New products are designed so that they can be produced in the most efficient way. Products already in production are constantly evaluated. Changes may be made in a product's design to decrease production time.

Fig. 4-16. These efficiency engineers are studying the operations used to manufacture farm equipment.

Time management is a science. Production managers in both manufacturing and construction must continue to look for new ways to make products faster. This must be done without harming quality. Companies will then be able to compete better in world markets.

► ► ► **FOR DISCUSSION** ◄ ◄ ◄

1. How would you prefer to be paid, by the job or by the hour? Why?
2. How could your own use of time be made more productive? Give an example.

Construction

Materials, machines, and labor must be organized to keep construction projects on schedule. If time is lost, the cost increases. Project managers organize the project. The manager schedules deliveries and hires workers.

Detailed records must be kept of the time needed for each part of the project. When one stage is nearly finished, materials and workers for the next stage must be ready. Fig. 4-16. If materials are not at the site when needed, the workers' time is wasted. Costs will increase. The project will fall behind schedule.

Research is being done to make construction more efficient. In residential construction, large sections of houses already made in factories are being used. Sometimes the entire home is built in a factory. The homes are shipped by truck to the site. A crane lifts them onto the foundation. This process reduces time and costs.

Extension Activity

■ Conduct a time and motion study of your morning between waking up and arriving at school. Keep an accurate record of how long it takes you to brush your teeth, eat your breakfast, walk to the bus stop, and so on. Then analyze your movements. How many steps between bathroom and bedroom? Between home and bus stop? In what ways could you reduce the movements and time it takes you to get ready for school by ten minutes?

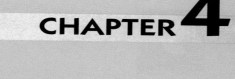
Chapter Highlights

● Information is an important resource of production systems.

● Sources of information include product histories, market research, laws and regulations, production schedules, and feedback.

● Energy is the ability to do work. Its sources are either unlimited, renewable, or limited.

● There are six forms of energy: light, heat, mechanical, electrical, chemical, and nuclear.

● Potential energy is stored energy. Kinetic energy is energy in motion.

● Time must be managed by production systems. Time and motion studies use scientific principles to improve production.

Test Your Knowledge

1. What organization was established in 1970 to protect the safety of workers?

2. What agency enforces laws protecting the environment?

3. Where did all energy on earth originally come from?

4. Name the three fossil fuels.

5. What kind of energy is uranium used for?

6. List the six forms of energy.

7. What is energy in motion called?

8. When a single energy source is used for two purposes, what is it called?

9. What is another name for scientific management?

10. Who first developed motion studies?

Correlations

LANGUAGE ARTS

1. In the library, use the *Reader's Guide to Periodical Literature* to find articles about the Chernobyl nuclear reactor accident in the Soviet Union. Prepare a report telling who, what, when, where, why, and how.

SOCIAL STUDIES

1. As a class, study the ways unlimited, limited, or renewable resources have been used throughout history by people of different cultures.

SCIENCE

1. Carve a cavelike hollow in a large block of styrofoam to create a solar cooker. Line the hollow with small mirrors. Take the cooker outside and use it to toast marshmallows.

MATH

1. You can do an assembly job alone in 60 minutes. If a classmate helps, you can complete the job in 30 minutes. How long would it take if a third person joined you?

CHAPTER 5

Materials

Introduction....................................

As you know, materials are one of the seven resources for production systems. Since early times, humans have used materials to make things. In fact, major periods in history have been named after the most important material of that time.

During the Stone Age, which began about one million B.C., stones were used to make tools and weapons. Fig. 5-1. Then metals became important during the Bronze Age. Bronze is created by combining two metals: copper and tin. Then iron took the place of bronze. Today iron is still important in the making of steel.

What material do you think our own time period could be named for? Some people suggest the Silicon Age. Silicon is used to make computer chips!

In this chapter you will learn about the different materials used in production today.

After studying this chapter, you should be able to......................

Discuss the most common materials used for production and their properties.

Tell how materials are located and processed.

Discuss the difference between standard and specialty stock and tell what each is used for.

List the questions designers and engineers ask when choosing materials.

Discuss the advantages and disadvantages of the just-in-time system of delivery.

Words you will need ....

alloys
composites
corrosion
hardwoods
inventory
just-in-time (JIT) delivery
lead time
mechanical properties

primary processes
sensory properties
smelting
softwoods
synthetic materials
thermoplastics
thermoset plastics

Stone Age	Bronze Age	Iron Age	Silicon Age?
500,000 B.C. - 3,000 B.C.	3,000 B.C. - 1,500 B.C.	1,500 B.C. - The Present	The Present -The Future

Fig. 5-1. The early periods of human history have been named for materials developed at the time.

Materials Used for Production

Both natural and synthetic materials are used in production. Natural materials are found in nature. They include such things as clay, wood, metal ores, water, air, coal, oil, and natural gas.

Materials created by people are called **synthetic materials**. They are products of technology. Rubber is an example. Natural rubber comes from rubber trees but is no longer used much. Synthetic rubber was developed and produced in a laboratory from a mixture of chemicals.

The primary materials used in production systems include metals, woods, plastics, ceramics, and composites.

Metals

Metals are found in the earth in substances called metal ores. The ores are mined and then broken down so the metal can be removed. This separation of the metal from ore is called **smelting**.

The resulting metals are natural, or pure, metals. Nothing has been added to them.

Few natural metals are strong enough to be used in industry. Natural metals may be made stronger by combining them with another metal or other substances. These combinations are called **alloys**. Fig. 5-2.

Metals commonly used for production include:
Aluminum. This lightweight metal conducts heat and electricity and resists rust. It is often used in alloys. Products include lawn furniture, airplane parts, and appliances.
Zinc. Natural zinc is used to create a coating that resists rust. This coating is used on steel and other metals. Zinc is alloyed with copper to create brass. Brass is used in doorknobs and plumbing fixtures. When zinc is combined with copper and nickel it forms nickel silver, which is used to make jewelry.
Copper. By itself copper is a good conductor of electricity. It also resists corrosion. (**Corrosion** is the wearing away of a material by acid or rust.) For these reasons, it is used to make electrical wire and water pipes. Copper is alloyed to make brass, nickel silver, and bronze (copper and tin).

Fig. 5-2. What alloys do you recognize among these items?

Lead. Lead is heavy and soft. Batteries are made from lead. It resists corrosion. Lead is combined with tin to make solder. Solder is then used to join metals together. It is used in plumbing and electronics.

Tin. Tin resists rust. For this reason it is used to coat the cans in which food is sold. Tin is also used to create bronze and solder alloys.

Chromium. Bright, shiny chromium is used to coat other metals. It resists rust and is attractive. Many products, such as automobile bumpers, are coated with chromium. Stainless steel is an alloy containing chromium.

Iron. Pure iron is seldom used. This is because it rusts too easily. To harden iron, it is combined with small amounts of carbon. One alloy formed by this is **steel**. Steel is one of the strongest alloys and has many uses. Tools, wire, nails, and beams are all made of steel.

Woods

Woods do not all look the same. Some may be pale in color and have many knotholes. Others may be dark brown or red and have a fine grain. They also differ in hardness and strength.

Hardwoods come from trees that lose their leaves in the fall. Oaks and maples are examples. Hardwoods tend to be hard, strong, and expensive. Fig. 5-3, Part A. They are used for furniture, floors, and sporting goods, among other things.

Softwoods come from cone bearing trees, such as pine and redwood. These trees usually keep their needles, or leaves, all year long. Softwoods are not as hard or strong as hardwoods, but they are still useful. Fig. 5-3, Part B. They are used in house construction. Doors, shingles, and framing are all made of softwoods. They are also used for such things as furniture and shelving.

Some wood is simply cut to size for lumber. Other wood is processed into wood products. These wood products have many uses in manufacturing.

Plywood. Plywood is made of thin layers of wood laid at right angles to one another. These layers are glued or cemented together. Plywood is strong and solid. It is used for many things, such as furniture and cabinets.

Hardboard. Wood fibers are pressed and glued together to make hardboard. It is not as strong as plywood. Hardboard is used for paneling and in some furniture.

Particleboard. Wood chips and other wood waste is mixed with glue and pressed into particleboard. Particleboard is used to make shelves and furniture.

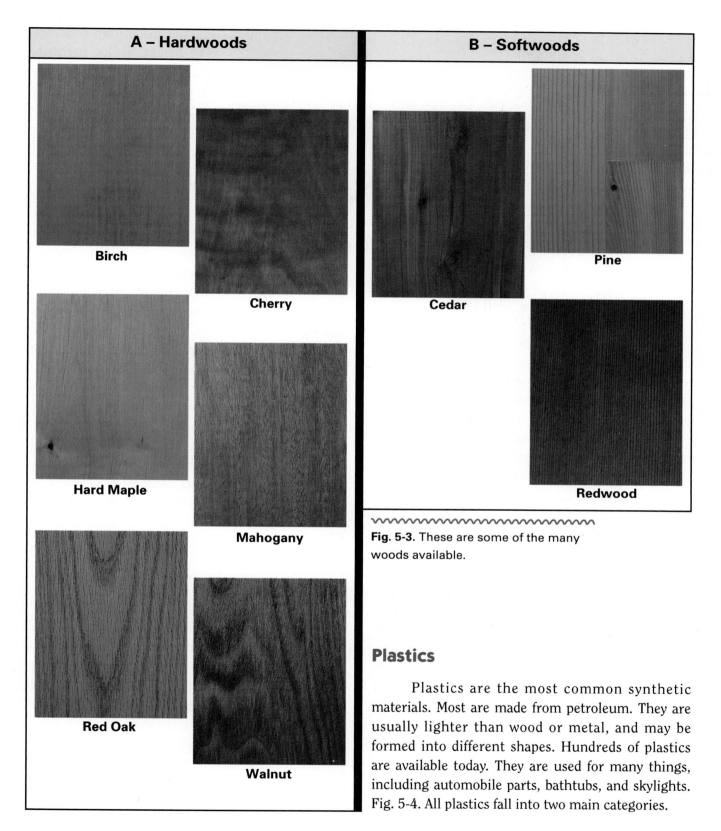

Fig. 5-3. These are some of the many woods available.

Plastics

Plastics are the most common synthetic materials. Most are made from petroleum. They are usually lighter than wood or metal, and may be formed into different shapes. Hundreds of plastics are available today. They are used for many things, including automobile parts, bathtubs, and skylights. Fig. 5-4. All plastics fall into two main categories.

Fig. 5-4. These are only a few of the thousands of items made from plastics.

Thermoplastics are heated until they are soft and then molded into shape. Later, they can be reheated and remolded. As a result, thermoplastics can be recycled. They are used in such things as food wraps, trash bags, toys, and telephones.

Thermoset plastics can be heated and shaped only once. They remain "set" in that shape. If they are reheated they may burn. Thermoset plastics are used in such things as car bodies and dinnerware.

Ceramics

For the most part, ceramics are synthetic materials made from clay, sand, and similar substances. Heat is required to make them. Most ceramics cannot be recycled. Glass is the exception. It can be reheated and reshaped into a new product.

Ceramics are strong and hard, and they resist corrosion. They also make good insulators for heat and electric current. (An insulator prevents something such as heat from passing through.) This is one reason ceramic tiles were chosen for the outside of the Space Shuttle. Ceramics are also used for cookware. Certain ones can go from freezer to oven without breaking. Auto manufacturers are testing engines made from ceramics. The bricks, sinks, bathtub, and windows in a house are usually made of ceramic materials.

Composites

Composites are made of two or more materials. The materials themselves are not changed. Each brings its own qualities to the composite. One material acts like a glue holding the others together. Sometimes one of the materials is actually glue. Plywood is a composite made from sheets of wood and glue. Concrete is another widely used composite. It is made from cement, sand, gravel, and water.

Composites are used to make tennis rackets, lightweight bicycle frames, and fishing rods. The body of the new B-2 bomber aircraft is partly made from carbon/graphite fiber composites. The use of composites in the bomber's design makes it hard to detect on enemy radar. Fig. 5-5.

Other Materials

Production systems use many other materials. One is textiles. Textiles are fibers woven together to make fabrics. Wool and cotton are natural fibers. Wool comes from sheep. Cotton is grown as a plant. Polyester is a synthetic material made from petroleum. Textiles are used to make many products, such as clothing, carpeting, rope, and parachutes. Fig. 5-6.

Leather is made from animal skins. A major use of leather is for shoes. It is also used in some clothing and for wallets and handbags.

Fig. 5-5. The body of the B-2 Stealth bomber is made from composites.

Fig. 5-6. In a textile mill, giant looms weave fibers into fabrics.

Both plant and animal materials are used in food manufacturing. Grains, such as wheat, are used for flours and cereals. Vegetables, such as soybeans and sugar beets, are turned into oils or sugars.

Finishing materials are used to decorate or protect other materials. Paint is a finishing material for wood. So are wax, lacquer, and varnish. Many of these are made from petroleum.

▶▶▶ FOR DISCUSSION ◀◀◀

1. **Plastics are made from petroleum, a limited resource. Yet many plastic items, such as trash bags, are disposable. This means they are meant to be thrown away. Do you think this is an efficient use of petroleum? Why or why not?**

2. **Name a product that can be made from either natural or synthetic materials. Give one advantage and one disadvantage of using each.**

3. **Should researchers continue to develop new synthetic materials? Why or why not?**

Extension Activity

■ **Collect as many raw materials, such as clay, wood, metal, textiles, and sand, as you can. Using them, create a class display and label each one. Make drawings of some of the products using these materials.**

Properties of Materials

Materials have many different properties, or characteristics. These properties determine whether a material can be used for certain production needs. Some properties can be judged by using our senses. Others must be tested.

Sensory Properties

Sensory properties are those we react to with our five senses — sight, hearing, taste, touch, and smell. Some of these properties include:

Color. Color can be pleasing to the eye. It can also be used to call our attention to something. Sometimes a product's color can have a special meaning. Stop signs, for example, are always red.

Texture. The texture of a material indicates how rough or smooth it is. Texture is important to appearance. It can also affect how a product is used. Kitchen appliances, for example, usually have a slick surface. This makes them easier to clean. Sofas, on the other hand, are made with soft materials. This makes them more comfortable.

Temperature. Some items are made from materials that can help keep us either warm or cool. Clothing, for example, can keep us warm.

Odor. Some materials are used for their pleasant odors. The substances used to make perfumes are an example. Have you ever bought popcorn in a movie theatre because it smelled so good? Odors are essential to manufacturers of many consumer items.

Flavor. Also important to food manufacturers is the way a material tastes. People are fond of certain flavors, such as chocolate. Flavor may also be important when making such products as mouthwash.

Sound. Some materials are used for the special sounds they produce. The material used to make guitar strings is an example. Other materials are used because they can deaden sound. Special insulating materials may be put within walls to keep noise out.

Mechanical Properties

Mechanical properties have to do with how a material reacts to loads and forces. Fig. 5-7. Testing can determine mechanical properties. By means of testing, the best material for a product can be chosen.

If a material has high ductility, its shape can be changed

without breaking it. Aluminum is very ductile. This is why it is used in foils, such as food wraps, that are meant to be shaped or folded.

Compressive strength is the opposite of tensile strength. It is the ability of a material to resist being crushed. The foundation of a building must have compressive strength.

If a material is elastic, it can be pulled or pushed out of shape, but it will return to its original shape. Rubber is an elastic material.

A material has plasticity if it will keep its new shape after it has been deformed. Modeling clay is a plastic material.

Shearing is pushing a material in opposite

directions at the same time. This is what happens when you tear a sheet of paper in half. You pull one half toward you and the other half away from you.

When you're fatigued, you're tired. Fatigue strength is the ability to carry a load over and over again without wearing out. The metal

used to make airplanes must have fatigue strength. Otherwise, over time the plane will develop fatigue cracks.

When a force is applied to brittle materials, they break easily. Window glass is brittle.

A hard material cannot be scratched or dented easily.

Diamonds are the hardest natural material. They must be cut with special tools.

Torsion is a kind of twisting action. A material that has high torsion strength cannot be twisted easily.

Impact strength is the ability of a material to withstand a

Tensile strength is the ability of a material to resist being pulled apart. Materials with tensile strength are usually very dense or compact. If a rope used for tug-of-war has little tensile strength, it will break.

sudden blow. If you ride your bike into a brick wall, you will discover the wall has impact strength.

Fig. 5-7. A material's reaction to force and loads determines its mechanical properties.

Other Properties

Materials may also be useful because of other properties. Some of these include:

Chemical properties. These have to do with how a material reacts to chemical compounds. Corrosion resistance is a chemical property. This tells how well a material resists the chemical reactions caused by air and water. Iron will corrode, but aluminum will not. Many metals are painted to increase their corrosion resistance.

Optical properties. A material's optical properties have to do with how it reacts to light. Opaque materials do not allow light to pass through. Some opaque materials, such as aluminum foil, reflect light. Translucent materials allow some light to pass. A sheet of white paper is translucent. If you use it to cover a window, it will not keep out all the light. Transparent materials admit all the light. Clear window glass is transparent.

Electrical properties. Materials that allow electricity to pass through easily are called conductors. Copper and gold are good conductors. Superconductors are made from special materials cooled in liquid nitrogen. They conduct electricity better than any other material. Insulators block the flow of electricity. Rubber and plastic are good insulators. That is why they are used to cover electric wires. Semiconductors are neither good conductors nor good insulators. They are somewhere in between. Semiconductors have important uses in computers and electronic devices.

Thermal properties. How does a material react to heat? Those are its thermal properties. Most metals conduct heat well. They are used in radiators. Fiberglass does not conduct heat. It is used in buildings for insulation.

Magnetic properties. Some materials are attracted to magnets. Iron is an example. Other materials, such as aluminum, do not react to magnets at all. The superconductors mentioned above can be used to produce very strong magnetic fields. This allows for the easy flow of electricity. The magnetic levitation (Maglev) train being developed in Japan uses superconducting magnets.

▶▶▶ FOR DISCUSSION ◀◀◀

1. **What properties do you think materials for the following products should have? cake mix, denim jeans, child's swing, light bulb, toolshed, railroad bridge.**

2. **Look around the classroom. Pick an object and try to identify one of the materials from which it is made. What properties does the material have? Why are these properties important in the object you chose?**

Extension Activity

■ **Develop simple testing devices to test for the different mechanical properties. Use these devices to test common materials you find in the lab. Assign the materials a rating as to how well they do. Keep records of the results.**

IMPACT

Sometimes construction materials can cause health problems. Asbestos is an example. Asbestos is a fiber-like mineral once used for insulation. Eventually it was learned that asbestos caused serious lung diseases, including cancer. Thousands of workers became ill. It is now against the law to use asbestos as a building material.

Creating Industrial Materials

Raw materials must be located and processed before they can be used by production systems. Today, this may include hundreds of steps and involve workers from all over the world.

Finding the Materials

Sometimes, locating raw materials is easy. A farmer who plants a field of cotton knows exactly where to find that resource. Other raw materials may be more difficult to locate and gather.

Mining

Coal and metals, such as copper and iron, are mined from the ground. Open pit mines are used when the material is close to the surface. The top layer of soil is removed. Heavy equipment removes the coal or ore. Afterward, the land can be restored by filling in the hole. Underground mines are used for materials deep below the surface. Fig. 5-8. Even with modern safety methods, mining underground can be dangerous. Poison gases can be trapped in tunnels. Fires can break out. Tunnel walls can collapse.

Harvesting

Trees and other plant resources are harvested. Heavy equipment may be used to cut and process wheat and other crops. Trees are cut down with chain saws and taken to lumber mills.

Drilling

Oil and natural gas are located by drilling below the earth's surface. Drilling may be done on land or at sea from huge rigs. Fig. 5-9. Pipes are installed and natural pressure brings the material to the surface. Sometimes pumps must also be used. Barges and pipelines transport the material to refineries for processing.

Fig. 5-8. Mining is done for such things as metals and coal.

Fig. 5-9. This platform is in Alaska's Cook Inlet and is being used to tap a large natural gas field.

Fig. 5-10. In this sawmill timber is cut to size.

Processing

Many raw materials cannot be used by production systems until they are processed. During processing, the shape, size, appearance, or strength of the material may be changed.

Primary processes are those that change raw materials into industrial materials. There are three kinds of primary processes: chemical, thermal, and mechanical. Chemical processes involve chemical reactions. Thermal processes use heat. Mechanical processes include such actions as cutting and crushing.

When logs are processed, they are taken to a sawmill and measured. This determines how much lumber may be cut from them. In some mills, computers are used for more precise measuring. The greatest amount of usable lumber must be obtained from each log. The bark is removed by machines and then large saws cut the logs into lumber. Smaller saws cut the lumber to commonly used lengths. Fig. 5-10. These are all mechanical processes.

After this the lumber is then graded according to quality and dried. Drying helps control its moisture content. The lumber may be air-dried outdoors in sheds or heated in ovens called kilns. Heating the lumber is a thermal process. Kiln-dried wood is preferred for construction and furniture. It is less likely to warp or crack.

Metal ores from mines are taken to smelters for processing. The ore may be crushed, washed, and then heated to concentrate the metal. Iron ore, for example, is mixed with coke (from coal) and limestone. This is a chemical process. The mixture is then heated in a large furnace. The limestone mixes with any impurities to form slag. Slag is a waste product and is removed. The remaining metal is called pig iron. If the pig iron is used in an alloy, such as steel, it will go through more processing. Fig. 5-11.

Crude oil is processed in refineries and separated into different vapors or gases. The oil is heated. The resulting vapors are collected. These vapors are then condensed and form different products, such as heating oil, kerosene, and gasoline.

All these industrial materials are turned into finished products by means of secondary processes. The materials are reshaped, cut, combined with other materials, or given new properties. These processes will be discussed in Chapter 9.

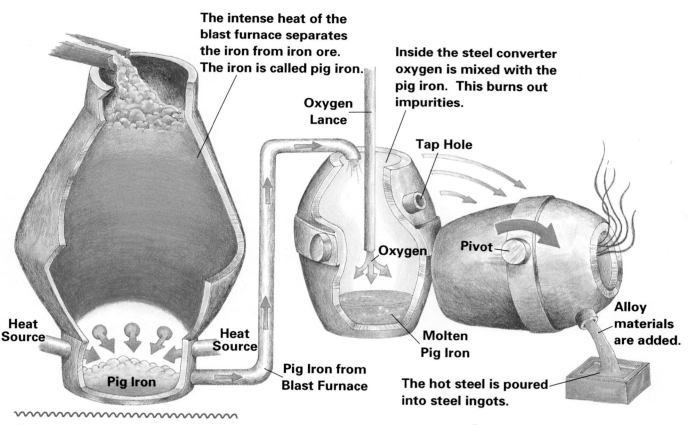

The intense heat of the blast furnace separates the iron from iron ore. The iron is called pig iron.

Oxygen Lance

Inside the steel converter oxygen is mixed with the pig iron. This burns out impurities.

Tap Hole

Oxygen

Pivot

Heat Source

Heat Source

Pig Iron from Blast Furnace

Molten Pig Iron

Pig Iron

Alloy materials are added.

The hot steel is poured into steel ingots.

Fig. 5-11. Steel is an alloy made from metals and other substances.

Preparing Stock

During processing, materials are given certain sizes. Lumber, for example, is cut to different lengths. Material produced in the sizes most commonly used is called standard stock. Material in custom-made sizes is called specialty stock. For example, the owner of a house may want a special window design. Fig. 5-12. Regular sizes won't work. The builder must order custom parts.

In general, standard stock costs less than specialty stock. It is also more readily available. For these reasons, many production companies prefer to use it.

Recycling

Many natural materials can be recycled. Paper products are an example. The paper is reused to make another product. This helps to save trees. Recycling also reduces production costs. Recycled aluminum costs about half as much as aluminum that must first be mined. Recycling aluminum uses only 20 percent of the energy needed for making new aluminum.

At present, there is a shortage of recycling plants. Also, certain items are not recycled enough. For example, only one percent of plastics is recycled. However, the use of recycled materials for production is growing. Paper, glass, plastic, and metals can all be reused.

▶▶▶ **FOR DISCUSSION** ◀◀◀

1. **Many production materials come in standard sizes. What advantages do you think this has for manufacturing and construction companies?**
2. **Does your community have a recycling plant? If so, what material does it recycle?**

Fig. 5-12. A window having a special design, such as this one, must be custom made using specialty stock.

Extension Activity

■ **Design an advertisement that promotes the 3 Rs: *Reduce* (the use of limited materials), *Reuse* (find ways to use products again), and *Recycle*. Decide where the ad should be seen or heard.**

Buying and Storing Materials

After materials are processed, they are sold. Manufacturing and construction companies choose carefully from among them.

Choosing the Best

Designers and engineers consider the advantages and disadvantages of many materials. Then they choose the one best suited to the job. They ask questions such as these:

Is the material suitable for the product? Will it work? The properties of materials under consideration are studied.

How much does it cost? Materials cannot be too expensive. The company must still be able to make a profit. Consumers must be able to afford the products.

Will the material be available when it is needed? Such things as strikes and shortages are taken into account. Also, the amount of time needed to make certain materials must be considered.

Will it have to be shipped? Shipping costs can add to the price of materials. Moving materials over long distances can mean delays. For these reasons, materials are often bought locally.

Does the company have the tools and machines needed to handle the material? Using a new material may mean a change in production methods. New tools and machines may have to be bought. Sometimes a new material may be rejected because a company is not ready to use it.

Ordering

The person in charge of obtaining materials for a project is called the purchasing agent. This person makes sure the right amount of materials is delivered at the right time.

Suppliers bid on the orders. In other words, they offer to sell the materials for a certain price. The bids are checked to see if specifications for the materials are met. The purchasing agent must also be sure the supplier can meet delivery schedules. The lowest qualified bid is usually the one accepted.

In arranging for deliveries, there must be enough lead time. **Lead time** is the time between the placing of an order and the delivery of the materials. For example, suppose it takes a supplier two weeks to make or obtain a material. Then at least two weeks lead time must be allowed. A purchasing agent may allow extra lead time in case any unexpected problems should occur.

When shipments arrive, the materials are checked. The amounts must be right. The materials must match specifications.

Storage

The purchased materials need to be stored. Space must be set aside in a warehouse or on the construction site. Proper storage of material requires planning and organization.

Inventories of material must be kept. An **inventory** is a record of what materials are on hand. Management must always know if there is enough to keep production going.

Storing materials for long periods can cause problems. The materials may lose quality. Food items may spoil. Metals may rust. Theft can also be a problem. The cost of replacing damaged or stolen materials must be added to production costs.

To avoid these problems, a new method is being used. It is called the **just-in-time** (JIT) delivery system. Under JIT, materials are delivered by the supplier just in time to be used. Few items are kept on hand. For example, a day's supply of engines might be delivered each day to a car manufacturer. The next day another supply of engines would arrive. No large warehouse is needed to store them.

Many construction companies also use JIT for delivery of materials to the site. For instance, steel beams may be delivered on the day they will be installed. Concrete is mixed just before it is to be poured.

Under JIT, materials do not have the chance to lose quality. They cannot be damaged in storage. They are less likely to be lost or stolen. However, JIT may not always work as planned. What if the materials fail to arrive on time? Suppliers must guarantee delivery. Even then, shipping delays may occur. In case of labor problems, there may be shortages.

▶▶▶ FOR DISCUSSION ◀◀◀

1. If you were a purchasing agent, which would you prefer, JIT or warehousing items? Why? What steps would you take to avoid any problems with JIT deliveries?
2. Suppose you are a designer who has been assigned the job of designing storm shelters. The shelters will be used on tropical islands. They must hold up under a lot of rain and wind. What materials would you consider using? What questions would you ask about each material?

Company Activity

■ Suppose you are the purchasing agent for your student company. You have to obtain the lowest price for the materials you need to make your product. Write or call several area suppliers to determine prices. Compare the prices to determine the best buys.

Chapter Highlights

- Products are produced from natural and synthetic materials. Metals, woods, and plastics are among the primary materials used in production systems.
- Materials have many different properties. They include sensory, mechanical, chemical, optical, electrical, thermal, and magnetic.
- Locating raw materials is usually done by mining, harvesting, or drilling.
- Processing a raw material may change its shape, size, appearance, or strength. Many natural materials can be recycled.
- The proper storage of materials requires planning and organization.

Test Your Knowledge

1. What is the term for materials created by people?

2. When natural metals are combined with another metal or substance, what is the result called?

3. What is the separation of metal from ore called?

4. Which type of plastic cannot be reheated and remolded?

5. What is the term used for a material composed of two or more materials?

6. Name four sensory properties.

7. If a product cannot be twisted easily, what property does it have?

8. By what three methods are raw materials commonly found?

9. What is a material called when it comes in custom-made sizes?

10. Why don't JIT systems require large warehouses?

Correlations

LANGUAGE ARTS

1. Interview a local fast food restaurant manager to determine if and how the company recycles. Prepare a list of ten questions to ask during the interview.

SOCIAL STUDIES

1. Display a collection of raw materials in your classroom. Tell how you think the world would be different if one of these raw materials disappeared from the face of the earth.

SCIENCE

1. Many people recycle aluminum cans. Design a can crusher that will flatten the cans and make them easier to store.

MATH

1. Alloys are made up of several types of metals. How much copper is in 25 pounds of an alloy that contains 5% copper?

CHAPTER 6

Tools and Machines

Introduction......................................

Ever since early times, humans have used tools to help them do their work. Probably the first tools were stones used to cut or crush. Later, people learned to use metals to make tools. Then tools became more complex.

Today, tools are changing again because of computers. These new tools can do all kinds of jobs. Computers make it possible to operate giant machines in space. Computer-driven robots may be used one day to make repairs inside our bodies. Already one scientist has developed a motor that weighs less than a speck of dust!

In this chapter you'll learn about the tools and machines used by production systems. You'll find out what these tools and machines do and the kinds of power they use.

After studying this chapter, you should be able to......................

Discuss the different types of tools and machines used in production.

Know and obey the safety rules for working with tools and machines.

Identify common hand and power tools and describe their uses.

Discuss the different power sources used for power tools and machines.

Give examples of heavy equipment and tell how each is used.

Words you will need ....

computer-controlled
   tools and machines
cranes
electric power
excavators

hand tools
hydraulic power
laser
maintenance
pneumatic power

Using Tools and Machines for Production

Production systems use thousands of different tools and machines. The type used depends on the work that needs to be done.

Types of Tools and Machines

Tools and machines can be placed into groups based on the kind of work they do. Fig. 6-1. Production tools and machines are used for:

Measuring. Measuring tools are used to determine sizes. They include such items as rules and scales.

Laying out designs for products. Layout tools and machines are used by designers to draw objects. They help them draw straight lines, angles, and circles. They may also be used to mark patterns or lines on materials, such as wood or metal. Some measuring tools may be used for layout and vice versa.

Separating materials. Separating tools and machines are used to divide materials into specific sizes and shapes. Cutting, drilling, and sanding are all separating processes. A saw is an example of a separating tool.

Forming materials. Forming tools and machines are used to change the shape of materials. Materials that are formed are usually soft or can be softened. Bending machines form metal by bending it in a variety of angles.

Combining the parts. Combining tools and machines join the parts of a product together. This may be done with glue, hammer and nails, or by many other combining methods.

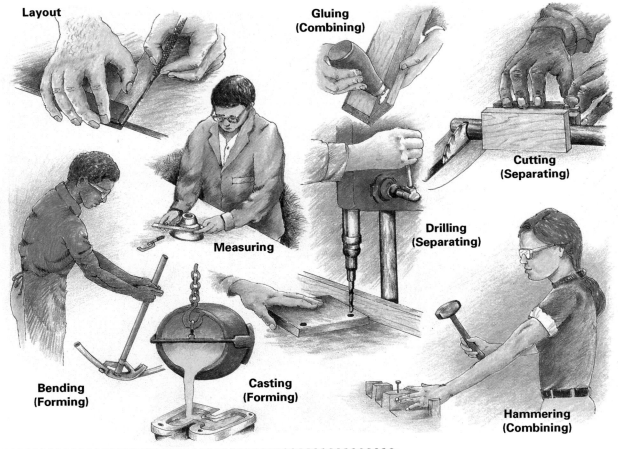

Fig. 6-1. Tools and machines fall into these basic groups.

Selection

The type of tool or machine used for a job depends on several things. The size of the company may be one. A small company often does not use the same kinds of equipment as a large company. Cost is usually an important factor.

The type of company is another. Tools and machines used for manufacturing are usually different from those used for construction.

The type and size of material being processed also affects the choice of equipment. Fig. 6-2. Finally, the type of operation and the safety and ease of using the equipment must be considered as well.

Safety

More than two million people are disabled in work accidents each year. Most of these accidents result from carelessness. Workers use tools or machines improperly. They fail to use the right safety equipment. They ignore the rules.

Industrial accidents affect everyone. Workers may be seriously injured. They may lose time and pay from work. The company's insurance costs may go up. This means higher production costs and higher prices for consumers.

Most accidents can be prevented. Workers need to learn the safe use of tools and machines. They must obey the rules that make a safer workplace for everyone. Fig. 6-3 shows you some general safety rules that should be followed.

Fig. 6-2. Sheet metal is cut with giant shearing machines. A factory processing sheet metal would probably require large areas in which to work.

▶▶▶ FOR DISCUSSION ◀◀◀

1. Accurate measurement may be more important for one kind of product than for another. Name some examples of products you think would require precise measurement. Give your reasons.

2. Look around your technology laboratory. Do you see any safety hazards? What should be done about them?

■ Think about the product your student company will be making. Make a list of safety rules that will apply during production. Post the list where everyone can see it. As a class, choose one person as safety inspector. That person will be responsible for pointing out any safety violations during production. He or she should keep records of violations. Each week's violations should be posted in the lab.

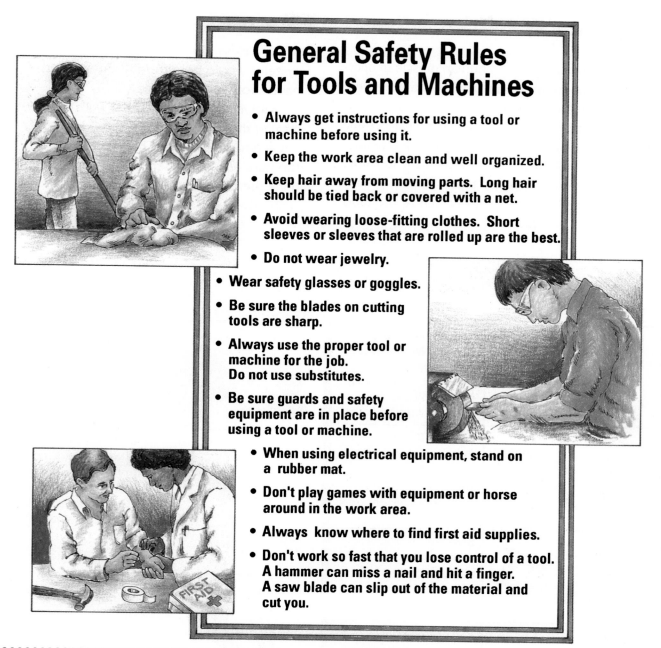

General Safety Rules for Tools and Machines

- Always get instructions for using a tool or machine before using it.

- Keep the work area clean and well organized.

- Keep hair away from moving parts. Long hair should be tied back or covered with a net.

- Avoid wearing loose-fitting clothes. Short sleeves or sleeves that are rolled up are the best.

- Do not wear jewelry.

- Wear safety glasses or goggles.

- Be sure the blades on cutting tools are sharp.

- Always use the proper tool or machine for the job. Do not use substitutes.

- Be sure guards and safety equipment are in place before using a tool or machine.

- When using electrical equipment, stand on a rubber mat.

- Don't play games with equipment or horse around in the work area.

- Always know where to find first aid supplies.

- Don't work so fast that you lose control of a tool. A hammer can miss a nail and hit a finger. A saw blade can slip out of the material and cut you.

Fig. 6-3. Read these safety rules and think about them. What is the reason behind each one?

Hand Tools

Hand tools depend on human muscle to operate. Many hand tools may already be familiar to you. You have used them at home or in the lab. Hammers and saws are examples. On the following pages some of the more common hand tools used in production are shown. Some, such as those used for forming, are seldom found in industry today. However, many school labs use them. Learn to recognize the tools and know their purpose.

▶▶▶ **FOR DISCUSSION** ◀◀◀

1. As a class, discuss how production systems would be different today if only hand tools were available.

2. With mass production, the skills for making things by hand are being lost. Only a few people still know how to make fine furniture and other items by hand. Do you think there is any point in preserving such skills? Give your reasons.

Adjustable open-end wrench. Wrenches are used to tighten nuts and bolts. This wrench can be adjusted to different sizes.	**Block Plane.** Planes smooth and trim wood by shaving off thin pieces. A block plane produces a fairly rough surface.
Aviation Snips. Aviation snips are used to cut thin sheet metal. The blades are very sharp and must be used with care.	**Box wrench.** This type of wrench slips down from the top and goes around hexagonal (six-sided) nuts and bolts. It is used to tighten them.
Backsaw. This saw has a reinforcing strip along its back. It is often used to cut wood molding.	
Ball peen hammer. A ball peen hammer is used to form both cold and heated metals. It can also be used to flatten certain bolts known as rivets.	**Claw hammer.** Hammers apply force to install fasteners, such as nails. The claw on this hammer may be used to remove nails as well.

Cold chisel. Cold chisels cut metals that have not been hardened. They are usually struck with a hammer. Care must be taken toward flying chips or sparks. The chisel should face away from one's body.

Compass. Compasses are used in laying out arcs and circles. They have sharp points and must be used with care.

Coping Saw. This saw has a narrow blade for cutting curves or irregular shapes.

Crosscut saw. These saws cut across the grain of the wood.

Flat file. Files have rows of teeth cut into their metal surfaces. When moved back and forth the teeth remove small amounts of wood, metal, plastic, and other materials. The larger the teeth, the faster material will be removed. The finer the teeth, the smoother the material's surface will be.

Flathead screwdriver. This screwdriver turns screws having a straight slot in the head.

Framing square. Squares are used to lay out angles. Framing squares are used mainly in construction by carpenters for angles having 90 degrees.

Hacksaw. A hacksaw is used to cut metal.

Half-round file. This file works the same as a flat file but is used on curved surfaces.

Hand drill. A drill is a rotating cutting tool that is pushed into a material to make holes.

Hand seamer. This hand tool is used to form hems and seams in sheet metal.

Level. A level determines whether or not a surface is level.

Locking pliers. Pliers are used to hold fasteners in place during installation. Locking pliers can be locked shut.

Long-nose pliers. Long-nose pliers are used to hold small fasteners in place.

Open-end wrench. This wrench slips around nuts and bolts from the side.

Phillips screwdriver. This screwdriver turns screws having a cross in the head.

Pipe bender. This tool is used in construction to shape pipes or tubes that protect wires and cables.

Ratchet wrench and sockets. The special handle on this wrench allows the worker to tighten a bolt without having to reposition the wrench after each turn. The socket fits over the nut or bolt during tightening.

Ripsaw. Ripsaws are used to cut wood along the grain.

Round file. This file works the same as a flat file but is used inside holes and other special jobs.

Rules. Rules measure distances. They may have divisions as small as 1/64th of an inch. Different kinds include steel rules and tape rules.

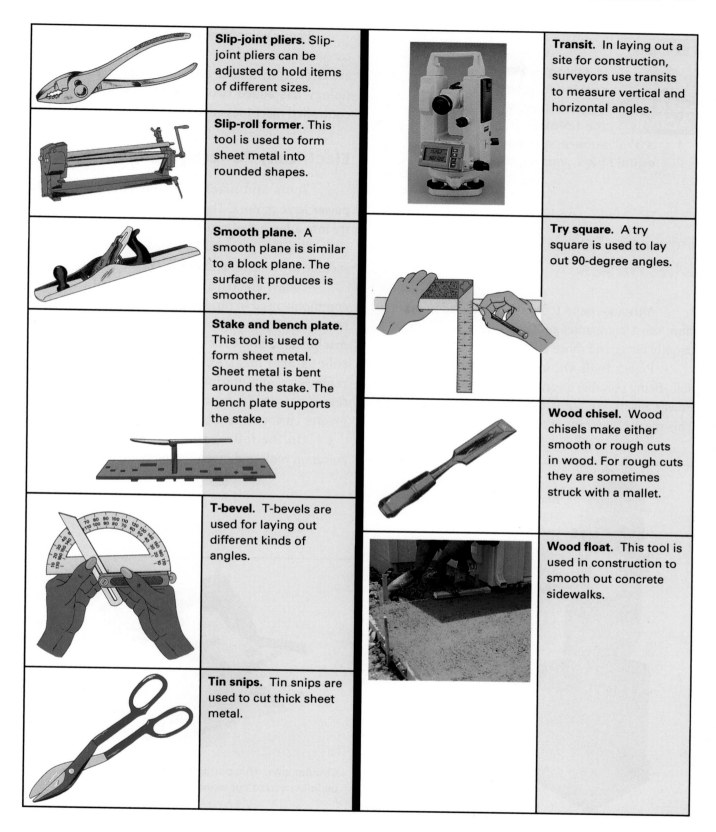

Slip-joint pliers. Slip-joint pliers can be adjusted to hold items of different sizes.

Slip-roll former. This tool is used to form sheet metal into rounded shapes.

Smooth plane. A smooth plane is similar to a block plane. The surface it produces is smoother.

Stake and bench plate. This tool is used to form sheet metal. Sheet metal is bent around the stake. The bench plate supports the stake.

T-bevel. T-bevels are used for laying out different kinds of angles.

Tin snips. Tin snips are used to cut thick sheet metal.

Transit. In laying out a site for construction, surveyors use transits to measure vertical and horizontal angles.

Try square. A try square is used to lay out 90-degree angles.

Wood chisel. Wood chisels make either smooth or rough cuts in wood. For rough cuts they are sometimes struck with a mallet.

Wood float. This tool is used in construction to smooth out concrete sidewalks.

■ **Using measurement and layout tools provided by your teacher, draw the following: a 2⅝" circle, a 3⁷⁄₁₆" square, a triangle with two equal sides, and an arch.**

Power Tools and Machines

Although hand tools are often used in production, most companies prefer power equipment. It is usually faster and more efficient.

Power tools are usually hand-held and portable. Being portable means they can be carried around. Power machines are usually larger and stationary. This means they cannot be moved easily.

Power equipment usually falls into three main categories based on the power used. These categories are electric, hydraulic, and pneumatic. Modern technology has added two new categories: computer-controlled machines and lasers.

Electrically Powered Equipment

Tools and machines that run on **electric power** have motors. The motor changes the electricity into a force that can do work. Many of these tools and machines are used to separate materials. They include saws, drills, and grinders.

Today, electrically powered robot arms are commonly used in production. One such arm is used to measure parts before and after they have been machined. A probe, or metal tip, on the end of the robot's arm measures how much material was removed. Computerized measuring systems mean fewer defects in products. Fewer defects mean greater customer satisfaction and higher profits.

On the following pages are shown several common tools and machines powered by electricity.

Band saw. Band saws are stationary and may be used to cut many materials.

Circular saw. This portable tool is usually used to cut wood.

Hand power drill. This portable drill will drill or bore holes in wood, metals, and other materials. Some used at construction sites are battery-powered.

Drill press. Drill presses are used in manufacturing to drill and bore holes in wood, metals, and other materials.

Jointer. Jointers trim wood surfaces and edges.

Planer. A planer is used to smooth wood surfaces.

Hacksaw. Power hacksaws are designed for cutting metal.

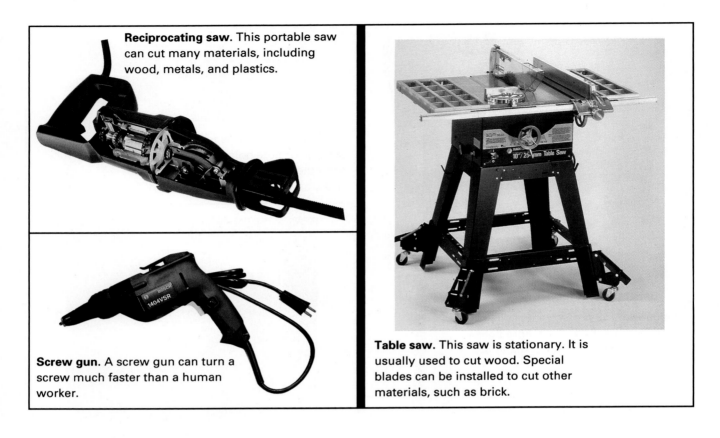

Reciprocating saw. This portable saw can cut many materials, including wood, metals, and plastics.

Screw gun. A screw gun can turn a screw much faster than a human worker.

Table saw. This saw is stationary. It is usually used to cut wood. Special blades can be installed to cut other materials, such as brick.

Hydraulically Powered Equipment

Hydraulic power comes from pressure applied to a liquid, usually oil. Pressure pushes the liquid through hoses to the area where the force is needed. Fig. 6-4. Hydraulic power is usually slow and steady.

Hydraulically powered machines are used in many areas of production. Forming metals is an example. Large hydraulic presses form such things as car fenders. In construction, backhoes and bulldozers rely on hydraulics. They are discussed under "Heavy Equipment."

Robotic arms are often controlled by hydraulics. Hydraulic pressure moves the arm up and down. It also opens and closes the grippers at the end of the arm. Robotic arms now do many of the manufacturing jobs humans used to do.

Pneumatically Powered Equipment

Pneumatics is similar to hydraulics. **Pneumatic power** comes from air pressure. Pressurized air is provided by an air compressor. The force of the air moves the parts of the tool or machine.

Air-operated machines can move more quickly than those using hydraulics. The steady, rapid force of a jackhammer comes from pneumatic power. Fig. 6-5. Air-powered nail guns also have quick-acting force. Their high speed helps drive the nail.

Manufacturing systems use pneumatic tools and machines to assemble products. Many portable tools, such as drills, screw guns, and ratchet wrenches, rely on air pressure.

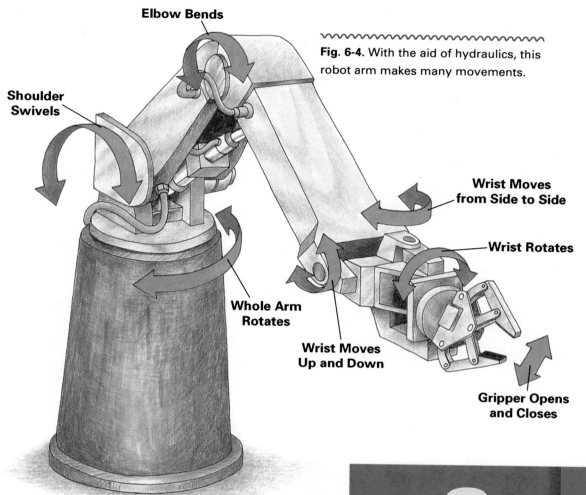

Elbow Bends

Shoulder Swivels

Fig. 6-4. With the aid of hydraulics, this robot arm makes many movements.

Wrist Moves from Side to Side

Wrist Rotates

Whole Arm Rotates

Wrist Moves Up and Down

Gripper Opens and Closes

Fig. 6-5. A jackhammer's power comes from air (pneumatic) pressure.

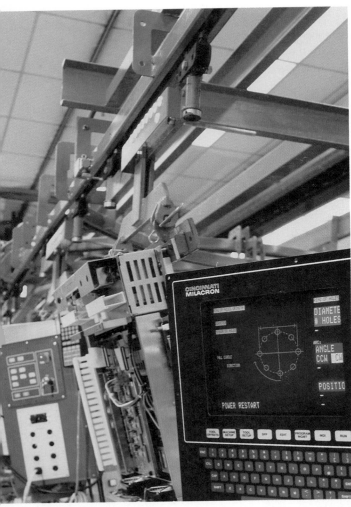

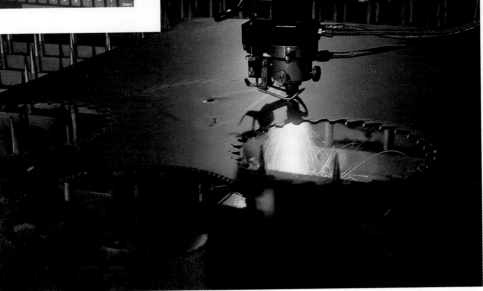

Fig. 6-6. By programming this computer, the machine can be told when, how, and where to move.

Computer-Controlled Equipment

Computer-controlled tools and machines may rely on electricity, air, or fluid as their power source. These power sources drive the moving parts. Computers, however, control how, when, and where they move. Fig. 6-6.

For example, a robot arm moves back and forth to pick up items on a conveyor belt. The power to move the arm comes from hydraulics. However, the computer tells the arm which items to pick up and where to put them.

Fig. 6-7. Because lasers are so powerful and the temperatures they reach are so high, they are controlled automatically.

Lasers

The word laser stands for **L**ight **A**mplification by **S**timulated **E**mission of **R**adiation. A **laser** is a narrow, intense beam of light.

Many uses for lasers are being found in construction and manufacturing. By linking a laser to a computer, the laser can be used to make extremely accurate measurements. Laser transits, for example, can measure angles very accurately.

Lasers are also being used to cut, drill, and bore materials. One laser can do the work of several ordinary cutting tools. Fig. 6-7. Because lasers reach such high temperatures, they can also weld materials together.

Use of lasers requires special safety measures. Looking directly into one or at its reflection can damage eyes. Lasers used for welding can cause burns.

▶▶▶ FOR DISCUSSION ◀◀◀

1. **Identify the power tools and machines in your technology laboratory. What power sources do they use?**
2. **In the lab, you are probably using power tools for certain operations. What hand tools have they replaced?**

Extension

Activity

■ **Suppose a piece of wood must be cut to 5" wide by 25" long. It must then be smoothed, and two ½" holes must be placed in it. Make a list of the tools and machines you would use for this work. Give reasons for your choices.**

Heavy Equipment

Machines used for construction are often referred to as heavy equipment. They do such jobs as moving earth or lifting heavy materials. Without them, many large buildings, highways, and other structures would be almost impossible to construct.

Cranes

Cranes can lift heavy loads and swing them from side to side. Most cranes used today are hydraulic. The load is attached to a long cable that is wound up on a pulley, called a hoist. A hook is used for general lifting. A bucket may be attached for carrying or digging.

Three commonly used cranes are shown in Fig. 6-8. A truck crane is mounted on a truck and can be driven from place to place. A crawler crane moves on metal treads over rough terrain. A tower crane, used for tall buildings, can be raised from floor to floor as the building goes up.

Excavators

Excavators are used for digging and moving earth from place to place. Almost every construction job requires some excavation. Most excavators are hydraulic. Several are shown in Fig. 6-9.

A backhoe is used for general digging. A bucket is attached to a supporting arm called a boom. The size of the hole dug depends on the size of the bucket. A trencher digs long narrow ditches for cables and pipelines. Front-end loaders are trucks with a large scoop at the front. They are usually used to carry small amounts, such as a load of dirt to a truck. A bulldozer is a tractor with a blade attached. Different blades are used for different purposes, such as to push earth around or clear land.

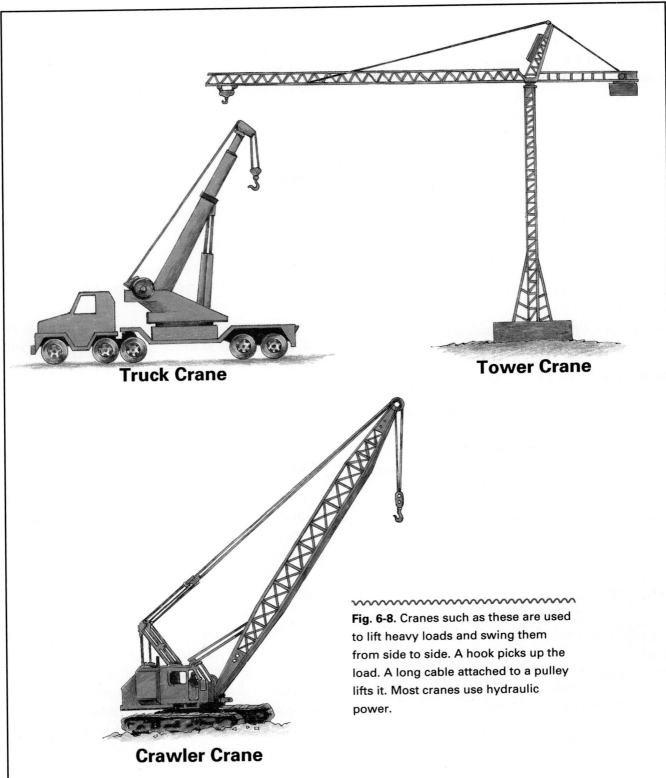

Truck Crane

Tower Crane

Crawler Crane

Fig. 6-8. Cranes such as these are used to lift heavy loads and swing them from side to side. A hook picks up the load. A long cable attached to a pulley lifts it. Most cranes use hydraulic power.

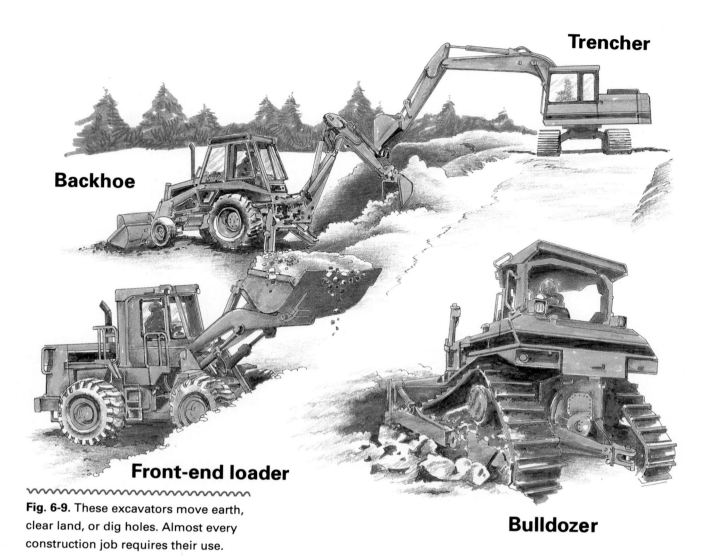

Trencher

Backhoe

Front-end loader

Fig. 6-9. These excavators move earth, clear land, or dig holes. Almost every construction job requires their use.

Bulldozer

IMPACT

One day certain natural resources, such as oil and gasoline, will run out. What will excavators and other heavy equipment use then for power? Makers of these machines must either invent a new kind of engine or be forced out of business.

Highway Construction Machines

Most highways are built using four types of machines — scrapers, graders, compactors, and pavers. Fig. 6-10. A scraper hauls soil from place to place. A grader levels the ground. A compactor rolls over the soil and packs it down. A paver spreads and smooths concrete or asphalt to create the road surface.

Fig. 6-10. Highways are built using heavy equipment. They prepare the land and lay the road surface.

▶▶▶ FOR DISCUSSION ◀◀◀

1. **What methods do you think early societies used to build large structures like the pyramids of Egypt? How do you think their methods differed from ours?**
2. **What effects do new highways have on towns and cities?**

Extension

Activity

■ 1. **Research the invention of heavy equipment. What was it first used for? How was it powered?**
■ 2. **Make a working model of a piece of heavy equipment. Use pulleys and string or syringes and tubing to create the movement that is required.**

Maintenance

Tools and machines are often expensive. They require a large investment of capital. If they must be replaced, they can reduce a company's profits. For this reason, they must be kept in good operating condition. Small repairs must be made before they become major problems. The regular servicing and repair of tools and machines is called **maintenance**.

Most companies have preventive, or scheduled, maintenance programs. This means that machines are checked regularly to keep them from breaking down. For example, on a construction site, the oil may be changed in a gas-powered engine after every 200 hours of operation. This prevents moving parts from wearing out too quickly.

Preventive maintenance takes time and costs money. However, it is a necessary part of production systems. The costs are small compared to the expense of repairs.

Large companies usually hire special maintenance workers who are trained to fix machines. Smaller companies may not be able to afford full-time help. Instead they may sign a contract with a local maintenance company. The maintenance company's employees are trained in machine repair.

Even with preventive maintenance, emergencies do occur. Many machine parts cannot be maintained regularly. Sometimes they simply break and the machine cannot be used. If parts are on hand, the broken ones can be replaced quickly. Many machine manufacturers offer emergency part service. Parts can be shipped by air within twenty-four hours.

▶▶▶ FOR DISCUSSION ◀◀◀

1. **Do the hand tools in your technology lab require maintenance? If so, what kind?**
2. **Do your parents arrange for preventive maintenance around your home? If so, what is it for, what must be done, and why is it done?**

Extension

Activity

■ **Make a schedule for preventive maintenance in the school lab. Include not only tools and machines but worktables and other equipment.**

Chapter Highlights

● Production systems use many different tools and machines. These tools and machines are used for all kinds of jobs, including measuring, laying out, separating, forming, and combining.

● Safety rules must be carefully followed to prevent accidents.

● Hand tools require human muscle to operate. Power equipment uses three main sources of power: electric, hydraulic, and pneumatic.

● Modern technology has led to such equipment as robots and lasers.

● Heavy equipment is used for such jobs as moving earth or lifting heavy materials.

● Regular maintenance keeps tools and machines in working order. Preventive maintenance tries to prevent problems before they occur.

Test Your Knowledge

1. What are forming tools and machines used for?

2. Name three types of wrenches.

3. What kind of hand saw is used to cut metal?

4. List the three main power sources for power equipment.

5. What liquid is usually used for hydraulic power?

6. What is the source of pneumatic power?

7. What is a laser?

8. Name three categories of heavy equipment.

9. What machine is basically a tractor with a blade attached?

10. What is the regular repair and servicing of tools and machines called?

Correlations

LANGUAGE ARTS

1. Acronyms are "words" created from the first letter of words in a long name or title. Using the dictionary, find the words that make up the following acronyms: LASER, SCUBA, RADAR, SONAR, NASA, OSHA, and COP.

SOCIAL STUDIES

1. As a class, discuss how tools and their uses have changed throughout history. How did the invention of these tools make lives easier for Americans during the Industrial Revolution?

SCIENCE

1. As you have probably learned in your science classes, there are six basic, or simple, machines. They are the lever, wheel and axle, pulley, inclined plane, jackscrew, and wedge. List one hand tool for each of the simple machines and tell which one it is based on.

MATH

1. A piece of metal measures $11\frac{5}{8}"$ in length. A piece $3\frac{3}{4}"$ long is cut off. How long is the remaining piece of metal?

You·Can·

— Make a Difference —

Students Get Things Rolling — Soon Everyone Is Recycling!

Start with 53 students, add a principal who's interested in the environment, and you soon have an entire community involved in recycling!

In 1989, Sister Ann Landers, C.S.J., became principal at St. Mary's School in Kickapoo, Illinois, a small rural community. Sister Ann talked to the school's 53 students about the importance of conserving natural resources. She pointed out that glass, plastic, tin, aluminum, and newspapers can all be recycled.

Students seemed interested, and Sister Ann made arrangements with a recycling firm. The firm agreed to send

a truck to the school parking lot the first of each month for a Recycle Weekend. Students invited everyone in the community to bring their waste to the truck. Many people participated. The money earned went to the school.

The following year newspapers were no longer taken by the recycling firm. However, the school continued to collect glass, plastic, tin, and aluminum...in the parish priest's garage! Every month, the students sorted things out, and volunteers with pickup trucks hauled the items to the recycling center.

The year after that, teachers at St. Mary's decided to add more about the envi-

ronment to what was already being taught in their classes. Students soon learned about everything from landfills to pollution to endangered animals. Through videos, talks, field trips, and discussions, they got involved in many special projects.

For example, they started compost buckets at school and taught their families how to make compost at home for yards and gardens. That summer they performed songs from "Gonna Have an Earth Day" at the 4-H booth at the local fair. They also calculated the amount of lunchtime trash they threw away on an "average" day. They compared that with the amount thrown away when they used lunchboxes and other reusable containers. They decided that, on the average days, many resources were being needlessly wasted.

Over the years, others in the community have joined St. Mary's students in their enthusiasm for saving the earth. Recently, a local medical center has begun supplying paper that's been used on one side to the school. Students use the paper as worksheets. This donation not only saves trees, it also saves the school money.

Can a small school with only 53 students make a difference? You bet. For its work on behalf of the environment, St. Mary's School/Parish has been named a "model" community by organizations in the area.

Like these students, you may want to start your own community recycling program. For information, contact your local county Cooperative Extension Office or
  GRDCA
    Attention: EPA Peer Match Program
    P.O. Box 7219
    Silver Spring, MD 20910
    Phone: 1-800-456-GRCD

MANUFACTURING

Activity Brief:
Manufacturing a Recycling Center

PART 1 : Here's the Situation...........

*I*n the late 1700s in Europe and America, a revolution began. Today, we refer to it as the Industrial Revolution. During the Industrial Revolution, products that had always been made by hand began to be made by machines. Factories came into being. The factories competed with one another for customers. Each tried to make the most products in the fastest time for the least amount of money. The age of modern manufacturing had begun.

In this section, you'll learn about different manufacturing systems. How do manufacturers decide what products to make? What processes do they use? How can they be sure their products are of good quality? These and many other questions will be answered.

PART 2: Your Challenge..........

Manufacturers today are faced with many of the same challenges as manufacturers during the Industrial Revolution. They must make large quantities of products in a very short time. They must pay attention to cost. They must make sure products meet quality standards.

In this activity, you will first design a product. The product will be a home recycling center that will help people organize their waste. Examples of centers are shown in Figs. A and B. Next, you will create a plant layout for manufacturing your product. Finally, you will organize a system of quality control to be used when making the product. Your teacher may also decide to have you manufacture the products in your own student "factory."

Design Tip:
For the ultimate in recycling, consider using recycled materials to make your recycling center. Sturdy cardboard containers and other items might work well.

A

B

PART 3: Specifications and Limits..........

Your work will need to meet certain standards. Read the following specifications and limits carefully before you begin.

1. Your recycling center design must be:
   - inexpensive to produce
   - easy to assemble
   - possible to manufacture in the school lab
   - used to hold and keep separate four kinds of waste: newspapers, aluminum, glass, and plastic

2. Your plant layout must also be suitable for use in the school factory.

3. You must hand in the following:
   - a log sheet of your work
   - rough sketches of at least 3 designs you considered

- a finished working drawing of the design you are submitting; anyone studying your drawing should be able to make the item
- a mock-up or prototype of your recycling center
- a drawing of how the plant must be laid out in order to manufacture your recycling center; all workstations must be labeled as to what will be done there
- a flow chart listing all the steps necessary in assembling your recycling center; all quality control checks must be indicated

4. You will be evaluated on the basis of:
   - originality
   - how well you researched the problem
   - how usable your solution is

1. Cut to Length
2. Counterbore Hole
3. Drill Large Hole
4. Cut Decorative Kerf
5. Machine Sand
6. Drill Small Holes
7. Hand Sand
8. Stain
INSP. Inspect
9. Finish
INSP. Inspect
10. Package
TS Hold for Distribution

PART 4: Materials...........

There are many materials you might use to design and build your recycling center. Here is a list of possibilities.

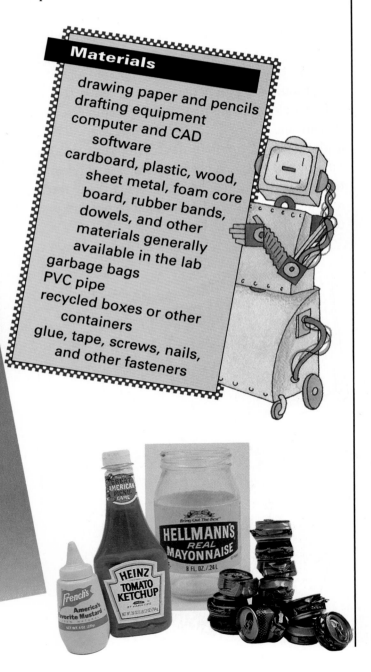

Materials

drawing paper and pencils
drafting equipment
computer and CAD software
cardboard, plastic, wood, sheet metal, foam core board, rubber bands, dowels, and other materials generally available in the lab
garbage bags
PVC pipe
recycled boxes or other containers
glue, tape, screws, nails, and other fasteners

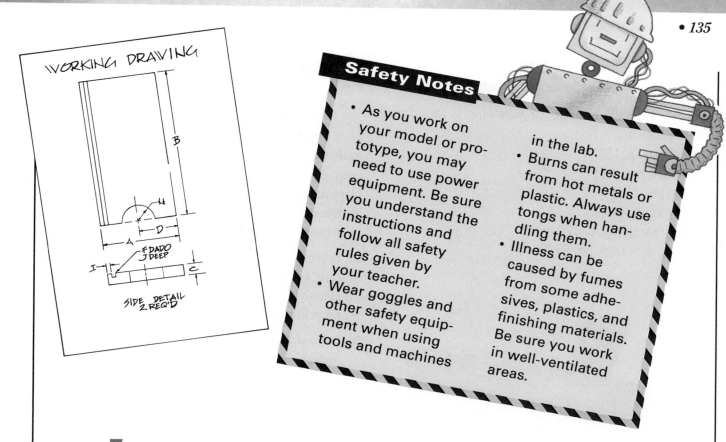

WORKING DRAWING

SIDE DETAIL
2 REQ'D

F DADO
J DEEP

Safety Notes

- As you work on your model or prototype, you may need to use power equipment. Be sure you understand the instructions and follow all safety rules given by your teacher.
- Wear goggles and other safety equipment when using tools and machines in the lab.
- Burns can result from hot metals or plastic. Always use tongs when handling them.
- Illness can be caused by fumes from some adhesives, plastics, and finishing materials. Be sure you work in well-ventilated areas.

PART 5: Procedures ....

1. Be sure to keep a daily log of your work.

2. Visit local stores or look through catalogs for ideas for your recycling center. At the same time, be thinking about what materials you could use. Keep in mind that a recycling center for home use should fit in a kitchen, basement, or garage. Think, too, about the sizes and shapes of the waste items that it will contain.

3. Make several sketches of designs that might be suitable. Remember that your design must be one your student company could make in the lab. Select the design you like best and submit it to your teacher for approval.

4. Make a finished working drawing of your design.

5. Make a mockup or prototype of your recycling center. Keep a list of the steps you take. It will help you when you make your flow chart.

6. Study the machines and layout of the school lab. Review the list of steps you used in making your model or prototype. Create a flow chart and plant layout for mass producing your recycling center.

7. Determine quality control standards. What would be considered a major defect? A minor defect?

PART **6**: For Additional Help...........

Information about the following is found on these pages in the text:

flow charts, p. 151
plant layout, p. 151
prototypes, p. 58
quality control, pp. 172-182
random sampling, p. 179
working drawings, pp. 56-58

PART **7**: How Well Did You Meet the Challenge?...........

When you've finished your recycling center, evaluate it. Does it do what it's supposed to do? Ask yourself the following questions. Your teacher and classmates may take part in this evaluation.

1. Does your recycling center provide space for newspapers, aluminum, glass, and plastic?

2. Is your working drawing complete? Does it contain all needed information, including dimensions? Using it, could someone easily make your recycling center?

3. In your flow chart, are all the steps in the proper order?

4. Is your plant layout organized for speed and efficiency?

5. How might you improve on your design? On your plant layout?

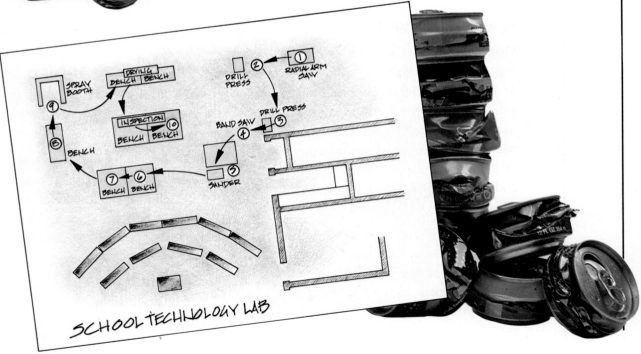

SCHOOL TECHNOLOGY LAB

Did You Know...?

• Every week, about 750,000 trees are used to produce newspapers. Two-thirds of those newspapers are never recycled. Paper takes up more than 40 percent of all landfill space.

• Americans go through about 2.5 million plastic bottles every hour.

• The average American throws away about 60 pounds of plastic every year. Plastics make up about 30 percent (by weight) of all household garbage.

• Paper towels made from recycled paper should not be used in microwave ovens. They have been know to catch fire when exposed to microwaves.

PART 8: Extending Your Experience..........

Think about the following questions and discuss them in class.

1. Does your family collect items for recycling? If not, why not? What do you think would persuade them to start?

2. At present, few communities have recycling plants. Why do you think this is so? What do you think would persuade more business people to open recycling plants?

3. Waste disposal and energy shortages are big problems for our society. Bioconversion (p. 84) is being used in some areas to solve both problems. In what other ways might waste be turned into energy?

CHAPTER 7

Getting to Know Manufacturing

Introduction....................................

What do you mean when you use the word "manufacture"? Most people use it to refer to products being made in a factory. That is what the term means for us today. It was not always true, however. Products were manufactured long before factories existed.

Until about 200 years ago, there were no factories. Skilled craftspeople manufactured products by hand. Each product, even those that were supposed to be alike, such as two shoes, was a little bit different. It was *unique* — one of a kind.

Today, manufacturing is quite different. In some companies it is still fairly simple. In others, it is very complex. Computers and other forms of technology have brought many changes to manufacturing at all levels.

After studying this chapter, you should be able to......................

Tell the differences between custom manufacturing, flexible manufacturing, and mass production.

Discuss the advantages and disadvantages of custom manufacturing and mass production.

Explain how computers are being used in manufacturing today.

Describe several manufacturing systems that use computers.

Words you will need ....

assembly line

computer-aided design (CAD)

computer-aided manufacturing (CAM)

computer-integrated manufacturing (CIM)

computer numerical control (CNC)

custom manufacturing

customized

flexible manufacturing system (FMS)

interchangeable parts

intermittent manufacturing

mass production

quality circle

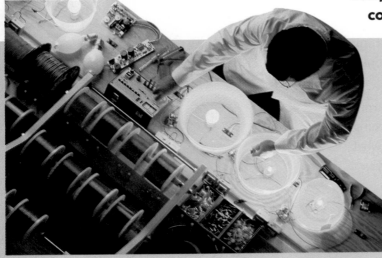

Custom Manufacturing

When products are specially made for customers in small quantities, it is called **custom manufacturing**. Many times only a single item is made. Industrial robots and artificial hearts are examples. Makers of custom products usually specialize in these products.

Very often buyers of custom products know exactly what they want. The product is then made to order. For example, the Navy may order a new battleship having special equipment. Fig. 7-1. Homeowners may ask that new kitchen cabinets be made to fit their own kitchens.

Sometimes a mass-produced item is **customized**. In other words, unique features may be added. For example, furniture can be customized. A customer may want a different covering, or upholstery, for a sofa. The manufacturer takes the mass-produced sofa and changes it to suit the customer. Fig. 7-2.

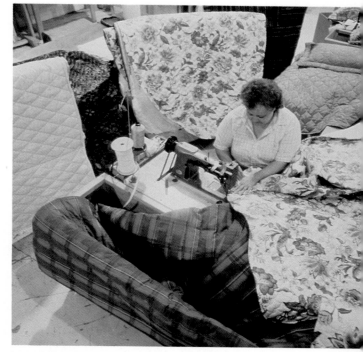

Fig. 7-2. This worker is sewing together pieces that will cover a sofa. Notice all the different fabrics that a customer could choose from.

Fig. 7-1. This ship, an aircraft carrier, is one of a kind.

Advantages and Disadvantages

Usually, custom-made products are built to higher standards. In addition, the customer gets a product made to order. The workers who make them are more involved in the whole production process. As a result, these workers often care more about their jobs.

There are, however, some disadvantages. Most custom products are expensive. This happens for several reasons. Because special materials may have to be bought in small quantities, they often cost more. The workers are highly skilled and able to do many tasks. This means they earn higher wages. Custom work also adds to labor costs because it takes longer than mass production.

▶▶▶ FOR DISCUSSION ◀◀◀

1. **If you worked in manufacturing, would you prefer making unique items or those that were all the same? Give your reasons.**

Extension Activity

■ **If you can, bring an item from home that was custom made. Try to find out where it was made and in what ways it is unique.**

Mass Production

Early automobiles, like the car shown in Fig. 7-3, were made using custom methods. All the parts were brought to one spot, and the car was assembled by hand. Each car took about twelve and one-half hours to complete. Then mass production was introduced to car manufacturing. Mass production methods were faster and more efficient. The 1927 *Model T* Ford shown in Fig. 7-3 took only 90 minutes to assemble!

In **mass production**, large quantities of a product are made on assembly lines using interchangeable parts. **Interchangeable parts** are

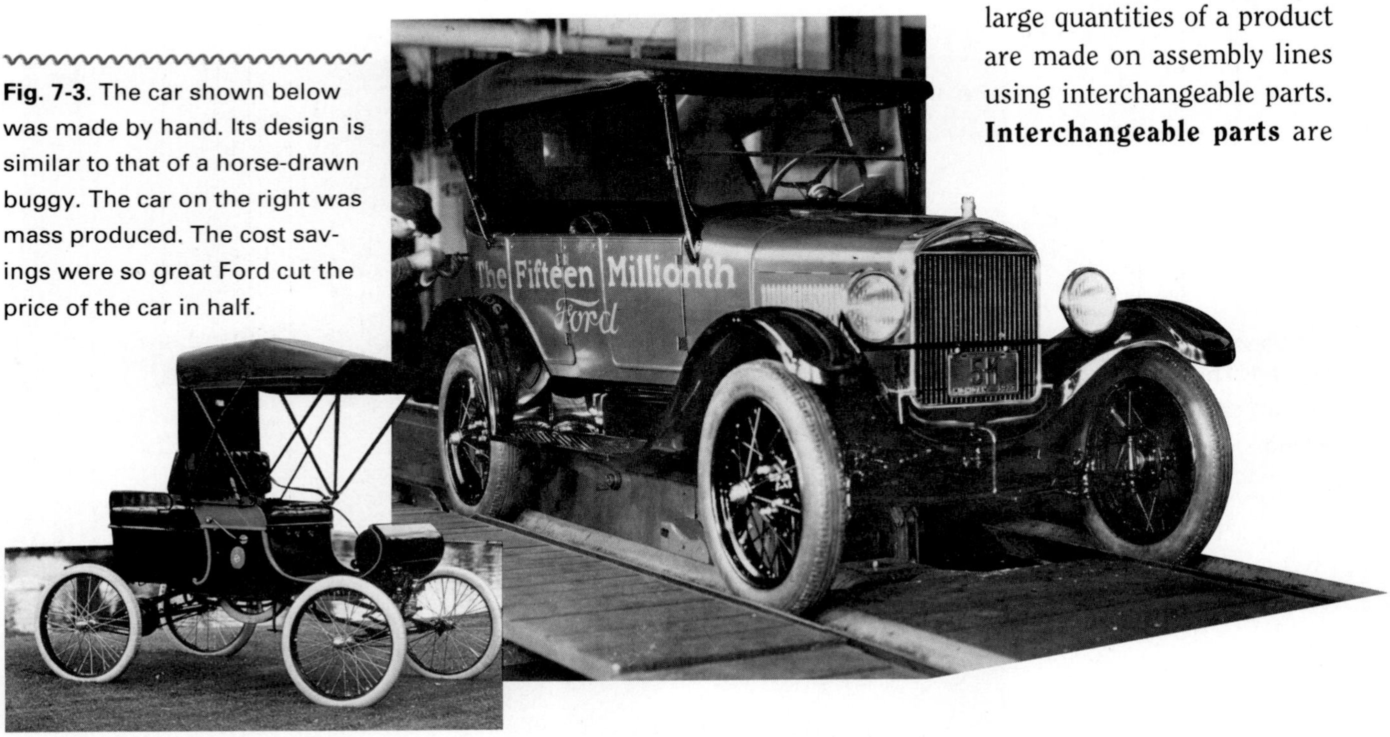

Fig. 7-3. The car shown below was made by hand. Its design is similar to that of a horse-drawn buggy. The car on the right was mass produced. The cost savings were so great Ford cut the price of the car in half.

Fig. 7-4. The interchangeable parts on the left can be used on any bike of the same design. On an assembly line, shown above, interchangeable parts make the work much faster. This assembly line is for tractors.

parts made exactly like one another. Any one of them will fit the product. Fig. 7-4. On an **assembly line**, the product moves past a line of workers. Fig. 7-4. Each worker adds a part, tightens a bolt, or does some other job. These tasks are carefully planned so that parts are put together in the proper order. When the product reaches the end of the line, it is finished. After the item is inspected, it is shipped from the factory.

Many products are now mass produced. They include cars, personal computers, clothing, and appliances. Sometimes, the workers stand on a conveyor belt and move alongside the product. In making small items, such as appliances, a factory may use several assembly lines. Each assembly line is fairly short. Also, computer-controlled robots are replacing human workers for many tasks.

It costs a lot of money to set up a factory to mass produce a product. Workers must be trained. Machines and other equipment must be adjusted or rebuilt. All these costs mean that large amounts of the product must be sold. The cost of the product would also go up if there were design changes each year. This is why some factories continue to use the same designs for many years.

If a product has many parts, its assembly line may have many subsystems. For example, a motorcycle production line has many assembly subsystems. The engine is produced on one assembly line. The frame is produced on another. Then they are put together on the main assembly line. Fig. 7-5.

Advantages and Disadvantages

Mass production has several advantages. It is fast and efficient. More products are turned out in less time. Most assembly line workers are unskilled or semi-skilled. They learn to do only one or two jobs. It costs little to train them, and often they are not highly paid. The company saves money.

Fig. 7-5. How many subsystems do you think a motorcycle would have?

A major disadvantage of mass production is that workers grow tired of doing the same job day after day. They feel little pride in their work. They may stop paying attention and become careless. Sometimes this results in accidents and products of lower quality.

Many manufacturers have tried to overcome this problem. Workers are retrained to do other tasks. They are moved to different areas of the factory and learn more about the product. They may be asked to join a quality circle. In a **quality circle**, workers meet to discuss production problems. Everyone offers suggestions for improvements. Fig. 7-6. Getting workers involved in decisions gives them more pride in their work. When this is done, the quality of products also improves.

Fig. 7-6. In a quality circle, everyone from managers to assembly workers is involved in making decisions.

▶▶▶ FOR DISCUSSION ◀◀◀

1. Many automobile parts fit only a certain make of car. Do you think that car makers should get together and use interchangeable parts that will fit any car? Why or why not?
2. Have you ever worked at a job you disliked because you were not included in decisions? What could have been done to make you happier with the situation?

Intermittent Manufacturing

During **intermittent manufacturing**, certain products are mass produced over a limited period of time. For example, every ten weeks an appliance manufacturer may change the production line to make a different product. During one period, lawn mowers are made. During the next, snow blowers are produced.

Most intermittent manufacturing systems use the same materials or machines to produce the different items. The assembly area remains basically the same from one product to the next. Wood furniture is often made this way. When production of coffee tables is completed, the factory is set up for chests of drawers.

Advantages and Disadvantages

An advantage of intermittent manufacturing is that the same equipment is used to make a number of different products. This saves money. The system is also flexible. If there is great demand for an item, production can be easily increased.

A disadvantage is that the factory must be set up differently for each new item. This can add to production costs.

▶▶▶ **FOR DISCUSSION** ◀◀◀

1. Wood and metal products cannot be made on the same production line. Why do you think this is true?

Company Activity

■ **Suppose your student company decided to make more than one product using the same materials and equipment. Determine what the second product should be. Then make a list of all the equipment and assembly line changes that would have to be made in order to make the new product.**

Computerized Manufacturing

Types of Systems

The computer is making many changes in manufacturing. Computer-aided design and manufacturing, computer numerical control, flexible manufacturing systems, and computer-integrated manufacturing all depend on computers.

Computer-Aided Design

In the past, product designs were drawn using paper and pencils and other drafting equipment. As changes in the design were made, the drawings had to be redone. This process could take days.

Today, products can be designed using computers. This is called **computer-aided design (CAD)**. Fig. 7-7. Special software (computer instructions) allows drafters to draw anything that used to be drawn on paper. Changes can be made quickly and easily. What used to take days may now take only hours.

Fig. 7-7. This artist uses no pencil or paper. When he's finished, another machine or a printer will reproduce the drawing.

Computer-Aided Manufacturing

When computers are used throughout production, it is called **computer-aided manufacturing (CAM)**. CAM reduces production costs by reducing the numbers of workers needed. Parts are machined more quickly and accurately.

CAD/CAM is the combination of computer-aided design and computer-aided manufacturing. Parts are designed on a computer. The design is then sent to the computers that control the machines. These computers "read" the design and tell the machine how to make the part. With CAD/CAM, computers talk to computers. This speeds up and improves the accuracy of the manufacturing process.

Computer Numerical Control

Parts for most products used to be made with machines operated by workers. The workers studied the product design and made all measurements with hand tools.

Today, with **computer numerical control (CNC)**, a computer controls a machine's operation. Once the material or part is in place, the computer tells the machine when and how to process it. Fig. 7-8. Measurements are more accurate and the work is done faster. The computer constantly monitors the work. Any adjustments are automatically made.

Fig. 7-8. Using the computer monitor, this worker is adjusting the machine to make parts automatically.

Flexible Manufacturing Systems

Using computers, different models of the same product can be produced on the same assembly line during the same production run. This is called a **flexible manufacturing system (FMS)**. For example, several models of a car with a similar design may need to be welded. Codes on the body of each car identify which model it is. As the car passes on the assembly line, a computer reads the code and tells the welding robot which welds to make.

In flexible manufacturing, workstations are called work "cells." Each work cell performs several different operations. For example, one cell may be programmed to weld, drill holes, and install screws. FMS speeds up manufacturing and increases accuracy. When a new part or design comes along, the computer program is simply adjusted.

Computer-Integrated Manufacturing

As you have seen, computers may be used to design a product, to control machines, and to link design and processing operations together. Computers are also used in other areas of a company. Records for payroll, sales, personnel, distribution, and accounting are also kept on computers. Many companies keep these different operations separate.

In **computer-integrated manufacturing (CIM)**, however, they are all linked together.

CIM requires a "mainframe" computer. A mainframe is a large computer to which many workstations can be attached. Fig. 7-9. Information from all areas of the company is stored in the computer. When one department wants to know what another is doing, it simply calls up the information needed.

In computer-aided manufacturing, every machine in production is run independently of the others. When CIM is used, however, the computer links all the machines. If one operation runs into a problem, the other operations are automatically adjusted. The work is more efficient. Fewer mistakes are made.

In Chapter 5 you learned about just-in-time (JIT) delivery methods. JIT is used with computer-integrated manufacturing. Because the computer controls all inventories, last-minute orders are more accurate. An up-to-date record of all materials and supplies is always available.

Because computer-integrated manufacturing has many advantages, its use will continue to grow. It reduces the chance of two different departments doing the same work. Information about the entire company and its operations is always available. Efficiency and profits are improved.

Advantages and Disadvantages

Computerized manufacturing has several advantages. Because computers are fast and efficient, they help reduce production costs. They also make it easier to change a product. For example, new car models are a little different from last year's models. Computers can help decide the best and cheapest ways to make the necessary changes. Also, with such computerized devices as robots, dangerous jobs can be done safely.

There are some disadvantages, however. Computerized manufacturing is very expensive to set up. Small firms may not be able to afford it. Computers require highly skilled workers to install and maintain them. The wages of these workers may be very high. Also, many employees lose their jobs when computers replace them.

Company Activity

■ Simulate a CIM operation for your student company. Use a computer to track materials costs, student labor costs, production time, inventory control, and product sales.

▶▶▶ FOR DISCUSSION ◀◀◀

1. Operations in many factories are now computer-controlled. Computers do many of the jobs workers used to do. Do you think a factory that operates without any human workers could exist? Why or why not?

2. How do CAM and CIM differ? What are the advantages of each?

Fig. 7-9. With CIM, all areas of a company are linked by means of a mainframe computer. What would each of the departments shown here need to know about the others?

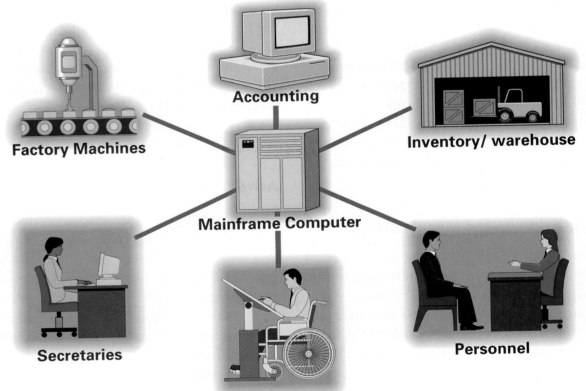

Chapter Highlights

● In custom manufacturing, products are made to order in small quantities.

● In mass production, products are made in large quantities on assembly lines.

● In intermittent manufacturing, a certain number of products are mass produced over a limited period of time.

● When computers aid in the design and manufacture of products, this is called CAD/CAM.

● When computers are used to control the operation of machines, this is referred to as cumputer numerical control (CNC).

● With flexible manufacturing (FMS), small quantities of similar items are made using the same computer-controlled machines.

● In computer-integrated manufacturing (CIM), all of the factory's computerized operations are linked together.

Test Your Knowledge

1. In what type of manufacturing are products made to order in small quantities?

2. Give two examples of custom products.

3. What are parts made exactly alike called?

4. During what type of manufacturing are a certain number of products made over a limited period of time?

5. What is the name of the group in which workers and managers meet to discuss production problems?

6. In what type of manufacturing is wood furniture often made?

7. What is the process called in which products are designed on computer?

8. What is computer numerical control used for?

9. In what type of manufacturing are all computer operations linked together?

10. What name is given a computer that has many workstations attached?

Correlations

LANGUAGE ARTS

1. Language is always changing to include new terms. "Laser" is an example. What other words based on technology can you think of that are probably new to our language?

SOCIAL STUDIES

1. Interview someone who works on an assembly line. Ask the person to describe the job. Does he/she feel an important part of the company? Why or why not?

SCIENCE

1. Design a display that shows how computers are used in automobile manufacturing.

MATH

1. A lawnmower manufacturer produces 55 lawnmowers per day. How many mowers will be produced in one 5-day work week? In one month?

CHAPTER **8**

Preparing for Manufacturing

Introduction.....................................

In Chapter 3 you learned how a company selects the products it manufactures. The next step is planning for production. Which is the best way to make the product? What kinds of machines are needed? Where should the machines be located? How many workers will it take to do the job? Only after these and other questions are answered does manufacturing actually begin.

After studying this chapter, you should be able to......................

Discuss the role of production engineering in making products.

List several questions engineers ask before choosing tools and machines for production.

Describe the three basic factory layouts.

Discuss what must be done in getting a factory ready for production.

Name several advantages and disadvantages in using robots.

Words you will need ....

conveyor belt
fixtures
jigs

production engineering
production quotas
subassemblies
tooling up

Forming

Forming is the process of changing the shape of a material without adding anything or taking anything away. Suppose you are working with a ball of modeling clay. Perhaps you flatten it out. Then you roll it into a coil. Since you have added nothing to it or taken nothing away, you have *formed* the clay.

Only melted or soft materials can be formed. Wire, sheet metal parts for cars, and water pipes are all shaped by forming processes. The processes include casting, compression, extrusion, drawing, stamping, and bending.

Casting

In casting, a material is melted or softened and then poured or forced into a mold. As the material becomes solid again, it takes the mold's shape. Both metals and plastics can be formed by means of casting. Fig. 9-1.

Metals

Sand casting is used to form metals. Wet sand is packed around a pattern that is shaped like the final product. Then the pattern is removed. Molten metal is poured into the empty space left by the pattern. After the metal cools, it is removed from the sand. It is now in the shape of the pattern. The molded item is called a casting. Fig. 9-2.

Any rough edges on the casting are sanded away. Sometimes holes must be drilled in it. Surfaces may also have to be machined flat and smooth. Because removing the metal destroys the sand mold, a new one must be made for each casting.

In permanent mold casting, a metal mold is used. The two halves of the mold are bolted together. Molten metal is then poured inside. When the metal has cooled, the mold is opened and the part is removed. The mold can then be reused.

Fig. 9-1. These items were all made using the casting process.

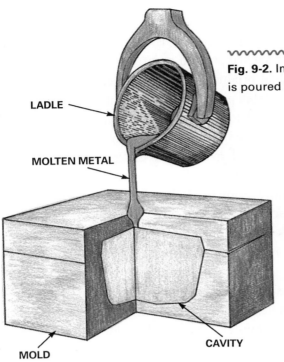

LADLE

MOLTEN METAL

MOLD

CAVITY

Fig. 9-2. In sand casting, molten metal is poured into a mold.

Plastics

To form plastics, methods called injection molding and blow molding are used. Fig. 9-3. The molds are made of two halves clamped together. In injection molding, plastic pellets are heated. When they form a soft mass, they are forced into the mold. To do this, a plunger using air pressure is used. The plastic hardens rapidly and is then removed from the mold.

In blow molding, compressed air is blown into a heated ball of plastic. The ball expands like a balloon against the inside of the mold. As it cools, it takes the mold's shape. Most plastic milk bottles and soft drink bottles are made this way. Fig. 9-4.

Fig. 9-3. These familiar items have been molded of plastic.

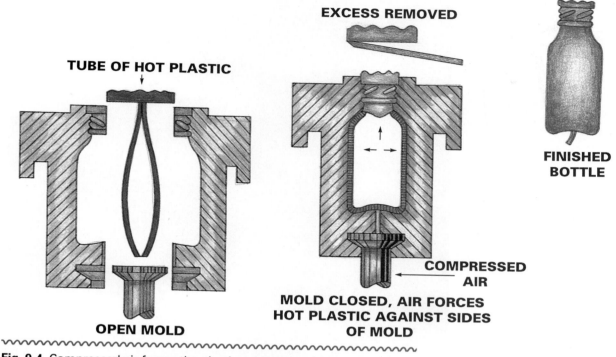

Fig. 9-4. Compressed air forces the plastic against the sides of the mold, leaving a hollow center.

Compression

Squeezing a material using force or pressure results in **compression**. Heavy or thick materials may have to be heated first, until they are soft and bend easily. This is called hot forming. Some thinner materials can be compressed at room temperature. This is called cold forming. The two most commonly used compression processes are forging and rolling.

You may have seen examples of forging in movie westerns. It was the method used by blacksmiths to form horseshoes. The blacksmith heated the metal until it was softened and then hammered it into shape.

In modern forging, dies are used to shape the metal. **Dies** are molds made of steel. Fig. 9-5. The metal is placed between two dies and compressed until it is the shape of the die. Hammer heads and wrenches are made this way.

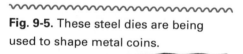

Fig. 9-5. These steel dies are being used to shape metal coins.

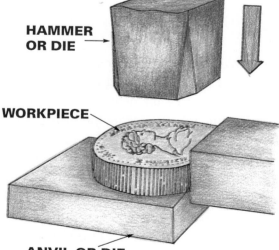

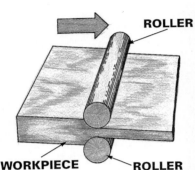

Fig. 9-6. This steel has been heated to soften it. Then it is rolled between rollers.

In rolling, the metal is squeezed between two rollers. Sometimes more than one set of rollers is used. As the metal is rolled, it gets thinner, longer, and wider. Sheet metal and steel beams are made by rolling. Fig. 9-6.

Extrusion

A material may be forced or pushed into its final shape. This process is called **extrusion**. The shape of the opening in a die determines the shape of the material. Fig. 9-7. Squeezing toothpaste from a tube is an example of extrusion. The toothpaste's shape depends on the opening at the top of the tube.

Soft metals, rubber, and some plastics can be extruded. Sometimes the materials are heated and sometimes they remain at room temperature. Almost any shape can be made. These shapes may be solid or hollow.

Fig. 9-7. Notice the shape of the die. The extruded material is formed into that shape.

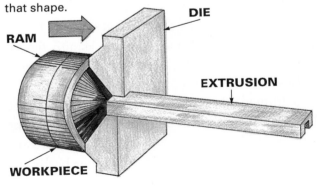

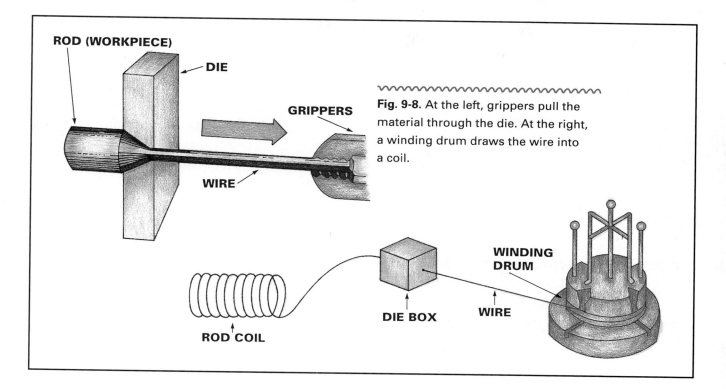

Fig. 9-8. At the left, grippers pull the material through the die. At the right, a winding drum draws the wire into a coil.

Drawing

Drawing is the opposite of extrusion. In drawing, the material is pulled, or stretched, through a series of dies until it becomes the desired size. Fig. 9-8. Wire is usually made this way.

Stamping

Stamping involves smashing the material between two dies. Sheet metal parts for cars are made by means of stamping. Fig. 9-9. If the final shape is to be complex, more than one stamping process may be required.

Fig. 9-9. These car parts have been made by the stamping process.

Bending

Copper tubing, thin strips of steel, and most sheet metal can be easily bent or twisted. During bending, one side of the material is compressed and the other side is stretched. Sometimes the material needs to be heated before bending. The type of material, its thickness, and how much it bends determine whether heating is necessary.

▶▶▶ FOR DISCUSSION ◀◀◀

1. **Look around the classroom. Try to identify products that have been formed. What method do you think was used?**
2. **Have you ever watched a blacksmith work? Describe what happened.**

Extension Activity

■ **When cookies are made with a cookie press, the extrusion process is often used. The dough is put into a tube and then forced through a die to create interesting shapes. Fig. 9-10. Design a die or series of dies for a cookie press. The shape created should be new and different.**

Separating

Separating is the process of cutting materials to size and shape. For example, you separate a material when you cut a piece of paper with a pair of scissors.

Materials have to be separated for many reasons. One important reason is to make them the correct size and shape for a product. Also, materials are easier to handle when reduced in size. Some material may be lost during separating.

Shearing

Shearing is cutting that is done with blades or dies by applying pressure. As the pressure increases,

Fig. 9-10. This shows cookies being pressed into shape.

the material fractures or tears along the cutting line. The resulting edge is usually smooth. No material is lost. Thin, soft materials, such as paper, fabrics, and sheet metal, are usually cut by shearing.

Chip Removal

Sometimes, cutting a material involves removing many pieces, or chips. The chips may be large, as when metal is turned on a lathe. The chips may be small and fine, as when wood is sanded smooth. Common chip removal processes include drilling, sawing, jointing, planing, routing, turning, milling, and abrasion.

Drilling

Drilling is done to make holes in materials. The drill bit, or cutting tip, is pointed and has sharp cutting edges. As it turns, it digs into the material. Fig. 9-11. Chips are removed as the hole is made.

Sawing

The sharp teeth on a saw blade must be harder than the material being cut. The teeth bite into the material and remove small chips. Wood, metals, and plastics can all be sawn.

Jointing

When wood is measured or cut, the results must be accurate. Jointing makes the edges of wood square and smooth so accurate cuts and measures can be made. Fig. 9-12. The jointer's three sharp cutters rotate as the wood passes over them. The cutters remove the chips of wood.

Planing

Planing levels and smooths wood surfaces. Sharp steel blades rotate as the wood passes over them. The wood is shaved off to make a flat and smooth surface.

Fig. 9-11. A drill bites into the material to make a hole.

Fig. 9-12. A jointer squares the edges of wood so that other processes will be done accurately.

Routing

During routing, a revolving bit removes material to shape the workpiece. Routing can be seen in the different types of edges on table tops. The grooves in kitchen cabinet doors are also created by routing. Routing is usually done on wood. Fig. 9-13.

Turning

Turning is done on workpieces shaped like cylinders. The workpiece is rotated against a sharp cutting tool. The tool removes chips. Many materials, such as wood, metals, and plastics can be turned to make such things as chair legs. Fig. 9-14. The machine on which turning is done is called a lathe.

Milling

Milling is a process used to produce flat surfaces, slots, gears, and keyholes. The cutters come in different shapes and sizes. They rotate as the material passes over them.

Abrasion

Abrasion occurs when chips of material are rubbed away with a harder material called an abrasive. The sharp fragments of abrasive bite into the material and remove it quickly. Fig. 9-15.

Grinding is also an abrasive process. Sharp, abrasive particles are glued to a wheel. Hard materials, such as brick, can be held against the wheel and ground into shape.

Fig. 9-13. A router was used to make the grooves in these items.

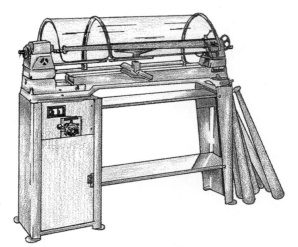

Fig. 9-14. The cutting tool on the lathe removes bits of wood to create the shape of these baseball bats.

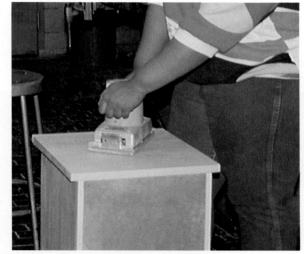

Fig. 9-15. The abrasive material on the sander cuts away tiny particles of wood.

Fig. 9-16. The water in this water jet cutter is traveling at a speed of 2,000 miles per hour. It can cut through metals and other materials that are up to 6 inches thick.

Non-Traditional Methods

Some new separating processes do not fit into the traditional groups. They include water jet cutting and various thermal and chemical methods.

Water Jet Cutting

In water jet cutting, water is pumped up to an extremely high pressure. It then travels through special hoses to a nozzle. In the nozzle is a tiny hole. The water is forced through the hole at speeds of about 2,000 mph. The tremendous force of this water jet enables it to cut through almost any material. Fig. 9-16.

Manufacturers are finding many uses for water jet cutting. It has several advantages. The water can be used over and over again. It never loses its sharpness. However, water jets can be dangerous for workers. For this reason, water jets are automatically controlled.

Oxyacetylene Cutting

Oxyacetylene cutting is a thermal separating process. **Thermal** processes are done with heat. The material is burned or melted away. In oxyacetylene cutting, two gases, oxygen and acetylene, are burned together in a torch similar to a welding torch. The flame from the torch heats metals until they are red hot and burn away. Fig. 9-17.

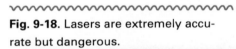

Fig. 9-17. The oxyacetylene flame burns the material along the cutting line.

Electrical Discharge Cutting

Electrical discharge machines (EDM) also do thermal separating. The cutting tool is first held over the material. Then the tool emits bursts of electricity. The electric bursts jump to the material and burn it away. EDM is very accurate and is used to make holes and curved surfaces.

Laser Cutting

Lasers can also be used for thermal separation of wood, metals, paper, and fabrics. As you know, lasers are narrow beams of light. The end of the beam becomes very hot. When focused on a material, the heat from the beam turns it into a vapor. Lasers are used in manufacturing when extreme accuracy and smooth surfaces are needed. Fig. 9-18. For safety and accuracy, they are automatically controlled.

Fig. 9-18. Lasers are extremely accurate but dangerous.

Etching

Etching is a chemical separating process. In chemical separating, chemicals eat into the material along a cutting line. For etching, acid is used on a metal workpiece. Electronic circuit boards, like those in computers, are etched.

▶▶▶ **FOR DISCUSSION** ◀◀◀

1. **Are any separating processes being used by your student company? If so, name them and describe what happens during separation.**

2. **Workers who separate materials often use gloves, goggles, or even masks. Why do you think this equipment is necessary?**

Extension

Activity

■ **Research and write a report on A) how lasers work or B) how water jets cut. Include diagrams, if possible. Also include a paragraph on the impact the laser or water jet has on manufacturing.**

Combining

Combining is the process of joining or adding materials together. Combining is used to mix materials, coat them, or assemble them.

Mixing

Mixing is the combining of two or more materials to form a new material. Mixing is commonly used by the chemical, food processing, and pharmaceutical industries. A maker of soup mixes, for example, combines several ingredients to make the product.

Coating

Coating is the use of one material to cover another. The material is covered in order to decorate or protect it, or both. For example, a car is coated with paint to protect the sheet metal from rust. The paint also makes the car attractive.

Plating is another kind of coating. A thin layer of one metal is used to plate another. Fig. 9-19. Many

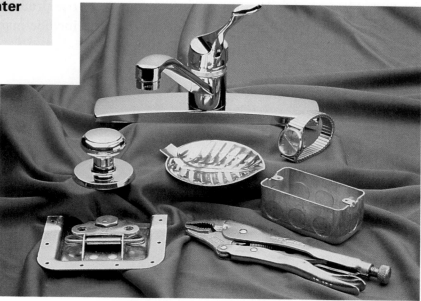

Fig. 9-19. These items have been coated. The coating protects them or makes them more attractive.

metal products are plated with chrome, a metal alloy. The chrome protects against rust and gives a shiny appearance.

Some products are coated with plastic. Tools are an example. Many window frames are also coated in this way. The plastic protects the wood frames from the weather.

Mechanical Fastening

When you drive a nail into a piece of wood to make something, you are doing **mechanical fastening**. A fastening device, such as a nail or a screw, is used to hold the pieces together.

Most mechanical fasteners are made of metal, although some are made from plastic or other materials. Nails, screws, staples, snap rings, cotter pins, clamps, plastic clips, rivets, and nuts and bolts are all commonly used fasteners. Fig. 9-20.

Adhesion

Adhesion is the use of adhesive, or glue, to hold parts together. Wood furniture, plywood, and laminated (multilayered) beams are all made by combining woods with adhesives. Plastic materials may also be combined in this way. Using adhesives

on metal, however, is a fairly new technology. The aircraft industry has done experiments with metal adhesives. Fig. 9-21. Adhesives are lighter and make assembly faster.

Soldering and brazing are also types of adhesion. In both processes, a material is melted and used to join two metal pieces together. The metal pieces themselves do not melt.

Cohesion

Cohesion occurs when two materials are melted together. Welding is a process based on cohesion. In welding, two pieces of metal are heated along the joint. The joint is the spot where they meet. When they melt, they flow together. As they cool and become solid, they form a tight bond.

Plastics can be combined using a dissolving liquid called a solvent. The solvent melts the plastic along the joint. The plastic then hardens again and forms a bond.

Assembly

When all the parts of a product are put together, the product has been assembled. Assembly may include many combining processes. Parts may be held together by mechanical fasteners, adhesives, or cohesion. Coating may take place either before or after assembly.

Fig. 9-20. Do you recognize these common mechanical fasteners?

▶▶▶ FOR DISCUSSION ◀◀◀

1. Review the different combining processes. What do you think would be some advantages and disadvantages of each?
2. In earlier times, people used nails, or pegs, made of wood as fasteners. Why do you think wooden fasteners are no longer common?

Extension Activity

■ Collect a number of different mechanical fasteners. Mount them on a piece of cardboard or create some other type of display. Label each fastener and list examples of what it is used for.

Fig. 9-21. Today, airplane manufacturers are using structural adhesives to put planes together.

IMPACT

In the past, manufacturers dumped dangerous wastes into waterways. It was thought that the waterways could absorb the pollutants. However, they could only absorb a certain amount. After years of this practice, the water became unfit for people and animals to use. In 1972, the Clean Water Act was put into effect by the government. The act now limits such dumping.

Conditioning

Conditioning involves changing the internal structure of a material. Its properties are improved by this. Some materials must be softened. Others must be hardened. Still others acquire stresses during processing that must be relieved. Some welds, for example, must receive treatments to remove strain caused by heat. Otherwise, the welds may crack. There are three types of conditioning: thermal, chemical, and mechanical.

Thermal Conditioning

In thermal conditioning, heat is used. Metals and ceramics often require thermal conditioning.

Heat is used to harden metals. This makes them resist wear better. Some are hardened only on the surface where the most wear occurs. This is called case hardening. Others must be hardened all the way through.

Hardening usually makes metals brittle. This means they will break easily. To remove the brittleness, they are heated again in a process called tempering.

Tempering makes metals tougher. Knives, chisels, and other cutting tools are often tempered. The blades stay sharp for a longer time.

Brittleness may also be removed by means of a process called annealing. Annealing makes metals softer. This is also necessary when metals become too hard during processing. Steel, for instance, becomes very hard when it is stretched and compressed. Annealing softens it again.

Chemical Conditioning

Adding chemicals to a material can also change its inner structure. This is called chemical conditioning. Rubber, for example, is conditioned when it is made into tires. The chemical process is called vulcanization. It makes the tires able to withstand heat better. Chemical conditioning is also used on many plastics.

Mechanical Conditioning

Mechanical conditioning involves using force on materials. Hammering a piece of metal against an anvil (iron block) conditions it mechanically. The molecules which make up the metal are changed. As a result, the metal hardens.

▶▶▶ **FOR DISCUSSION** ◀◀◀

1. **What do you think could happen if you tried to use the following items and they were not conditioned properly? 1) A hammer 2) A steel beam for a skyscraper 3) The gears on a machine**

Company Activity

■ **Determine the processes needed for your student company's product. Then set up a JIT delivery system. Schedule parts to be made as needed.**

Chapter Highlights

● Manufacturing processes fall into four basic groups: forming, separating, combining, and conditioning.

● Forming changes the shape of a material without adding anything or taking anything away. Methods of forming include casting, compression, extrusion, drawing, stamping, and bending.

● Separating is the cutting of materials to size and shape. The three categories of separating are shearing, chip removal, and non-traditional methods.

● Combining joins or adds materials together. Ways of combining include mixing, coating, mechanical fastening, adhesion, cohesion, and assembly.

● Conditioning changes the internal structure of a material. Types of conditioning include thermal, chemical, and mechanical.

Test Your Knowledge

1. Name the four groups of manufacturing processes.

2. What basic process changes the shape of a material without adding anything or taking anything away?

3. What is the term used for squeezing a material using force or pressure?

4. Name four common chip removal processes.

5. What separating process uses a very hot, narrow beam of light?

6. When chemicals are used for separating, what is the process called?

7. When materials are joined or added together, what is the basic process called?

8. Name three kinds of mechanical fasteners.

9. What is the name of the process in which materials are joined by melting them together?

10. Name the three types of conditioning.

Correlations

LANGUAGE ARTS

1. These words were used in this chapter. Without using a dictionary, write their definitions: molten (p. 157), machined (p. 157), fractures (p. 163), emits (p. 166), vapor (p. 166), solvent (p. 168), anvil (p. 170).

SOCIAL STUDIES

1. Write a short biography of the individual responsible for discovering one of the manufacturing processes mentioned in this chapter. Include the impact the invention had on other manufacturing processes.

SCIENCE

1. On a piece of paper, draw a design that will be etched on a piece of aluminum. Cut out the design. Wash and dry the metal. Lay the design cutout on the metal. Spray the metal around the design with a cooking oil spray. Remove the design cutout. Coat the design area with dilute hydrochloric acid (muriatic acid). Follow any safety tips from your teacher. When etching is complete, wash and dry the metal.

MATH

1. Find the diameter of a cylinder the circumference of which is 9.42 cm. (Use $C = \pi d$. Let $\pi = 3.14$.) Then measure the circumference of several manufactured items, such as a soft drink bottle and an aluminum can. Determine their diameters.

CHAPTER 10

Controlling Manufacturing Quality

Introduction......................................

Suppose you saw a commercial on TV about bicycles on sale at a popular store. The commercial promised that the bikes were of good quality. What would the words "good quality" mean to you in this case? Most likely you would expect the bikes to be well made. They should hold up well under normal use. The parts should fit snugly together and not wobble or rattle. The seat should be comfortable. The tires should last for several years. The paint should not be easily chipped or scratched.

Quality does not happen by accident. It is built into products. Quality requires an effort on the part of the manufacturer.

After studying this chapter, you should be able to......................

Discuss why quality is important.

Define tolerance.

Describe the two types of quality control programs.

Describe random sampling and tell why it is used.

Describe the three types of defects and give examples of each.

Words you will need ....

acceptance quality control
critical defects
gages
inspectors
major defects
minor defects
process quality control
quality assurance
random sampling
tolerance
warranty
zone of acceptance

Fig. 10-1. Many people are loyal to a certain brand of product because they have been happy with the company's quality standards.

Quality Can Reduce Costs

Initially, a quality product can be more expensive to make. The materials used may cost more. Time must be taken during production to make sure parts have no flaws. To do this work, **inspectors** must be hired. Fig. 10-2. However, attention to quality saves money in the long run. Fewer defective products will have to be thrown away. Additional production runs will not be needed to replace them. Workers will not have to spend time making repairs.

Fig. 10-2. Here, a team of workers inspects for quality.

Why Is Quality Important?

Quality assurance is the process of making sure products are made correctly. Most larger companies have an entire department to do this work. Why is quality so important to manufacturers? There are several reasons.

Quality Makes Customers Happy

Quality is important to customers. As a customer you know how it feels to buy a product that is poorly made. Customers want their money's worth. If they are happy with a product, they may buy others from the same manufacturer. Fig. 10-1. If they're unhappy, they will take their business elsewhere. Because of this, quality must be important to manufacturers, too. Manufacturers who lose too many customers cannot survive.

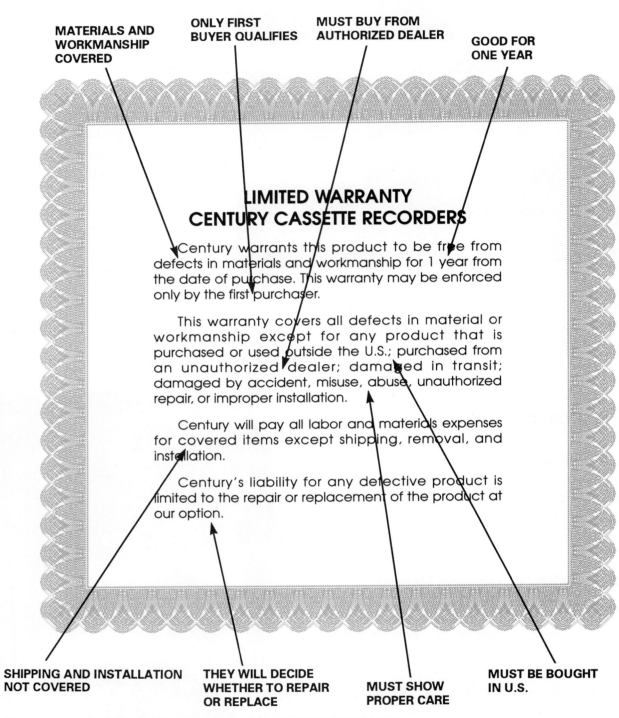

MATERIALS AND WORKMANSHIP COVERED

ONLY FIRST BUYER QUALIFIES

MUST BUY FROM AUTHORIZED DEALER

GOOD FOR ONE YEAR

LIMITED WARRANTY
CENTURY CASSETTE RECORDERS

Century warrants this product to be free from defects in materials and workmanship for 1 year from the date of purchase. This warranty may be enforced only by the first purchaser.

This warranty covers all defects in material or workmanship except for any product that is purchased or used outside the U.S.; purchased from an unauthorized dealer; damaged in transit; damaged by accident, misuse, abuse, unauthorized repair, or improper installation.

Century will pay all labor and materials expenses for covered items except shipping, removal, and installation.

Century's liability for any defective product is limited to the repair or replacement of the product at our option.

SHIPPING AND INSTALLATION NOT COVERED

THEY WILL DECIDE WHETHER TO REPAIR OR REPLACE

MUST SHOW PROPER CARE

MUST BE BOUGHT IN U.S.

Fig. 10-3. Look over this warranty. What exactly is the manufacturer promising?

Quality Cuts Down on Returns

Customers may return a product for many reasons. Most of those reasons have to do with quality. If the product is a food item, it may not taste good. If it's a clothing item, it may not fit properly. A sale has been lost. In some cases, the returned item may have to be thrown away. The manufacturer loses money.

To help protect customers, many products also come with a warranty. A **warranty** is a promise that the manufacturer will repair or replace the product if it does not work correctly. Fig. 10-3. Repairing or replacing is costly for the manufacturer. The better the product, the fewer returns and warranty claims.

Quality Means Fewer Legal Problems

Today, manufacturers are held responsible if their products have dangerous flaws. A manufacturer can be taken to court if a product causes injury or death. Fig. 10-4. Some companies have gone out of business because they have lost legal battles. However, the more attention given to quality and product safety, the fewer the legal problems.

IMPACT

Better transportation and a global economy have made goods available from all over the world. As a result, there is more competition among manufacturers. This competition means better values for consumers. For example, Americans liked the quality of Japanese cars. To compete with the Japanese, American manufacturers have improved the quality of their own cars.

How Good Is Good?

There are no perfect products. A product can always be improved. However, higher quality standards often mean higher production costs. That is why manufacturers must determine ahead of time how high the quality standards should be.

For most manufacturers, trade-offs have to be made. If standards are too high, the company may not be able to afford to make the product. If standards are too low, the customers may be unhappy.

Fig. 10-4. The position of the gas tank in the Ford *Pinto* caused it to explode in several accidents. Ford had to pay millions of dollars in damages.

The standards chosen are usually somewhere in between. Look at Fig. 10-5 and see if you can tell what sort of standards would be needed for the products shown.

Fig. 10-5. What quality standards would you suggest for each of these items?

Tolerance

Some products, such as scientific instruments, must be made to extremely accurate sizes. Even so, slight errors can still occur. How much larger or smaller than the requested size can a part be and still be used? The answer to that question determines the range of **tolerance**.

For example, suppose a part for the space shuttle must be 5 millimeters long. The tolerance is plus or minus half a millimeter. This means that the part can be as much as half a millimeter shorter or longer than 5 millimeters and still be used.

As they work, inspectors must consider tolerances. Parts that fall within tolerance limits are within the **zone of acceptance**. Parts that fall outside the zone of acceptance are rejected.

▶▶▶ FOR DISCUSSION ◀◀◀

1. **What personal characteristics should a good quality control inspector have? Give reasons for your answer.**

How Is Quality Checked?

The type of product usually determines the type of inspection made for quality. Some products, such as clothing, are simply looked over. Other products may have to be tested in a laboratory. Fig. 10-6. Cosmetics, foods, and drugstore items must be laboratory tested to be sure they are safe to use.

There are basically two types of quality control programs. They are acceptance quality control and process quality control.

Acceptance Quality Control

In **acceptance quality control,** inspections of parts are made at different times during production. The parts must meet specifications. If not, they are rejected before they go into a subassembly or into the finished product. Later, the subassemblies are also checked.

Finally, the finished product is also inspected. How does it work? Fig. 10-7. Does it do what it's supposed to do? Do the individual parts fit together properly? How does it look? Has the paint been applied correctly?

Process Quality Control

In **process quality control**, all parts of the manufacturing process are regularly inspected. This includes not only the products themselves but also the machines, methods, and workers.

Fig. 10-6. Laboratories are used to test many different products ranging from foods to shampoo.

Fig. 10-7. In this test, a refrigerator door is opened and closed many times to be sure it seals properly.

▶▶▶ FOR DISCUSSION ◀◀◀

1. **Which type of quality control program do you think would work best for your student company? Give your reasons.**

Company Activity

■ **Set up a quality control program for your student company. Develop a plan indicating when and how each part, tool, machine, process, or worker is inspected. Then "hire" inspectors to do the work.**

Even the best machines and tools wear out. To insure good performance, they are maintained and adjusted regularly. Replacements are made before the tool or machine begins to show poor results.

Quality is also maintained by keeping careful control over processes. For example, inspection may show that screws are loose on TV sets coming off an assembly line. Is the line moving too fast, so there is not enough time for the screws to be tightened? Are the parts not fitting together properly? By carefully checking the process, problems can be found and corrected. If the process is automated, changes can be made more quickly. Computer-controlled machines make constant adjustments throughout production.

Workers, too, are important to quality. In a process quality control program, every worker is also an inspector. If a worker sees a problem, he or she reports it to the supervisor. Some companies reward workers who discover problems or who think of better ways of doing things.

How Are Products Inspected?

Acceptance and process quality control use similar methods and tools.

Gages are measuring tools used to check sizes. Fig. 10-8. Lasers can also be used to check sizes and the position of parts. In an automated factory, computers control measuring instruments. As the parts are made, computers receive information and check dimensions.

Computers also perform other functions. They can check the operation of electrical circuits. Computer programs that analyze tool wear help assure that worn tools are replaced or repaired.

Fig. 10-8. A gage is being used to measure the diameters of these parts.

In mass production, only a certain number of products are inspected. A certain percentage of parts or products is selected at random. Fig. 10-10. Then those selected are tested for flaws or defects. Manufacturers calculate the number of flaws in the test sample. They then assume that the untested parts will have the same percentage of flaws. This technique is called **random sampling**.

Random sampling is used for both acceptance and process quality control. If too many defective parts are found, the production line is stopped. Inspectors find the reasons for the defects and correct them.

Random Sampling

In custom manufacturing, every product is inspected. For example, the parts of an oceangoing yacht are inspected as they are made. As the yacht is assembled, each part is tested for fit and appearance. Then the finished ship is taken out on the water to make sure it is seaworthy. Fig. 10-9.

Fig. 10-9. To be sure there are no leaks or other problems, this boat is being taken on a test sail.

Fig. 10-10. At this checkpoint, every hundredth grocery sack is checked to be sure the printing on it is clear and readable.

Extension Activity

■ Do a random sampling using colored marbles as "products." Place 89 red marbles and 11 green marbles in a container. The red marbles will represent products within the zone of acceptance. The green will represent those with flaws. Mix the marbles together. Cover your eyes and pick out 10 marbles. How many red and how many green are in your random sample? What is the percentage of green to red? If 10 percent of your production run could have defects, would you accept or reject the batch? Return all the marbles to the container. Repeat the experiment. Did the results change? What can you conclude from these experiments?

▶▶▶ FOR DISCUSSION ◀◀◀

1. As a customer, how do you feel about random sampling? Would you be willing to pay more for products if it meant each one was inspected? Give your reasons.

Company Activity

■ If your student company is manufacturing a product, do a random sampling during production. Decide what percentage of products you will check. Then check that number of products against the standards for the zone of acceptance.

What Do Inspectors Look For?

Inspectors look for three main types of defects in a product. These include critical defects, major defects, and minor defects.

Critical Defects

Critical defects can place the user of the product in danger. For example, a critical defect in an airplane engine could cause a fire, resulting in a crash. Fig. 10-11. Sometimes manufacturers are required by law to recall products that may cause serious injury or death. In a recall, the product is usually replaced or fixed in some way to make it safer.

Design or manufacturing errors are often the cause of critical defects. These errors must be found before too many items are produced and before consumers buy them.

Major Defects

A **major defect** can shorten a product's life span. It can cause it to fail during use. An example might be a shirt that is supposed to be washable by machine. If it falls apart when washed this way, there is a major defect. Major defects are not dangerous for the user. However, the product must be repaired or thrown away. Fig. 10-12.

Fig. 10-11. In what ways do you think each of these items could be critically defective?

Minor Defects

Minor defects do not affect a product's usefulness. They include such things as scratches, a mismatched color, or sags in a paint job. They usually pass quality inspections. Products with minor defects are sometimes marked and then sold as **seconds**. The customer then knows that the product has a defect. Fig. 10-13. The flaws are often difficult to notice. Seconds are usually sold for less.

DRAWSTRING NOT INSERTED

DYE MIXED IMPROPERLY

STITCHING UNEVEN

SLEEVES WRONG LENGTH

SEAM INCOMPLETE

Fig. 10-12. Which of the defects in this jacket would you consider major?

Fig. 10-13. Outlet stores, like this one, specialize in selling seconds.

▶▶▶ **FOR DISCUSSION** ◀◀◀

1. Suppose you are a quality control inspector in a plant that makes tires for cars. You have just discovered that the tires coming off the line have a critical defect. Heavy driving on hot days could cause them to blow out. However, the plant is behind schedule. Your supervisor does not want to shut down the line. She has told you to ignore the problem. What will you do in this situation?

Extension

Activity

■ Create a poster or classroom display showing the differences among the three types of defects.

Chapter Highlights

● Quality assurance is the process of making sure products are made correctly. Quality can reduce costs, cut down on returns, and prevent legal problems.

● The zone of acceptance includes products that fall within tolerance limits.

● The two types of quality control programs are acceptance quality control and process quality control.

● Gages are measuring tools used to check sizes.

● Random sampling is used to inspect products in mass production.

● Defects are divided into three categories: critical, major, and minor.

Test Your Knowledge

1. What is the process of making sure products are made correctly called?

2. What is the title of a worker responsible for product quality?

3. If customers are unhappy with the quality of a product, what do they do?

4. What is the term for a written promise to repair or replace a defective product?

5. What is the range called when parts are larger or smaller than the requested size but can still be used?

6. What are the two types of quality control programs?

7. What measuring tools are used to check sizes?

8. In mass production, what method is used to test a certain percentage of products?

9. What kinds of defects can place the user of the product in danger?

10. What are products called that will be sold with minor defects?

Correlations

LANGUAGE ARTS

1. Write a short story that has to do with some aspect of a critical, major, or minor defect in a product.

SOCIAL STUDIES

1. Write a brief report on the history of a product. Tell who invented it. Compare its quality when it was first put on the market to its quality now. Does a product always improve with time? Why or why not?

SCIENCE

1. Check the quality control of the weight of various products. For example, you might weigh large bags of candy, boxes of cereal, or bags of pretzels. How close is each product to the net weight written on the bag or box?

MATH

1. A part for a lawn mower must be 2½" long. The tolerance is plus or minus ¹⁄₁₆". How short could the part be in order to be used? How long?

CHAPTER 11

After Manufacturing

Introduction.....................................

The products have all been made. There they are, at the end of the assembly line, looking shiny and new. What happens next? Do manufacturers just sit back and wait for customers to come rushing in?

Not as a rule. Customers must first be told the product is available. A decision on how to ship the products to stores must be made. Will the product have a package? If so, it must be designed and made. All these tasks, as well as others, remain to be done.

After studying this chapter, you should be able to......................

Explain how products are priced.

Discuss the purpose of packaging and name several types.

List several ways in which products are advertised.

Explain the difference between wholesalers and retailers.

Explain why manufacturers maintain service centers.

Words you will need ....

bar codes	retailers
direct mailings	servicing
distribution	storage
marketing	warehouse
markup	wholesaler
packaging	

Pricing the Product

Before production begins, manufacturers usually have some idea of what they want to charge for a product. However, before production is complete, material costs may go up. The product may also take more or less time to make than estimated. These and other changes have to be taken into consideration. They may affect the final selling price.

The accounting department totals up all the costs involved. Not only production costs, but the costs of such things as research, advertising, and packaging must be included. Then the amount of profit desired is added. The total is divided by the number of individual products made. The resulting number is the price per item.

For example, suppose a run of 10,000 three-ring notebooks costs $23,400 to make. The manufacturer needs to earn $2,500 in profit. How much will the company charge for each notebook?

$23,400 + 2,500 = $25,900 ÷ 10,000 = $2.59

Manufacturers often suggest that stores sell the product at a certain price. This is called the manufacturer's suggested retail price (MSRP). Fig. 11-1. However, stores do not have to use this suggestion. The price is often determined by supply and demand. Some stores may raise it; others may lower it.

▶▶▶ FOR DISCUSSION ◀◀◀

1. What do you think would happen if a company failed to include all its costs when figuring the price of an item? How do you think such a mistake can be avoided?

■ **As a class, determine a selling price for your student company's product. First decide what percentage of profit you want for each item. Total all the costs involved, including the amount of profit desired. Then divide the total by the number of items produced.**

Fig. 11-1. Many retailers do not use the MSRP. Some omit it entirely. Others use it as a starting point for bargaining.

Packaging and Storage

The term **packaging** refers to any container that holds the product. This includes such things as boxes, cans, bottles, tubes, wooden crates, and plastic film. In most cases, the package is designed and made at the same time as the products themselves are. Many products are placed in the packages at the end of the assembly line. Then they are stored.

The Purpose of Packaging

There are several reasons why products have packages. Fig. 11-2. One reason is that some products need to be held in a container. Take milk, for example. How would you carry it home from the store if it were not contained in a bottle or carton?

Other products need to be protected from damage during storage or shipping. Eggs are one example. TV sets are another. If TV sets were not packed in cartons, how could they be stacked in trucks without being damaged? Packaging also protects certain items, such as food or medicines, from dirt and germs.

IMPACT

Another industry that has grown as a result of manufacturing is waste disposal. For some products, however, waste disposal is difficult. Some plastics, for example, do not decompose for hundreds of years. The average American throws away about 60 pounds of plastic each year. Plastics make up about 30 percent of the weight of all household garbage. Space in landfills is being used up.

Occasionally packages prevent shoplifting. A special package placed on audiotapes makes it harder for thieves to take them out of the store unseen. The package is too bulky to hide in a pocket or purse.

Packages are also used for advertising or to give information about the product. Bright colors and unusual designs attract customers to the package on the shelf. Some packages are designed so that the product is visible. This also is done to attract customers. What a package contains is usually printed somewhere on it. How to use the product is either printed on the package or on separate instructions inside.

In recent years, packages have been designed to protect consumers. Childproof medicine containers prevent children from opening them. Special seals on some items help warn consumers if the package has been opened. If the seal is broken, poisons or other harmful substances may have been added. Anyone who suspects a product has been tampered with should return it to the store unused.

Types of Packaging

The type of product determines the kind of package required. Some products do not need a package. These include lumber, steel beams, and pickup trucks. Products such as cereal and laundry detergent are placed in cardboard boxes. Glass or plastic bottles and metal cans hold liquids. Clear plastic "bubble" packs may be used for anything from cheese to children's toys.

For some items, the package keeps them separate. Hairbrushes, for example, are usually packaged as single units. For other items, the package helps group them together. This is why the pieces for a puzzle or a set of four folding chairs are boxed.

Sometimes small packages may be shipped inside still another container. These containers are called cases. Cases are easier to handle and ship.

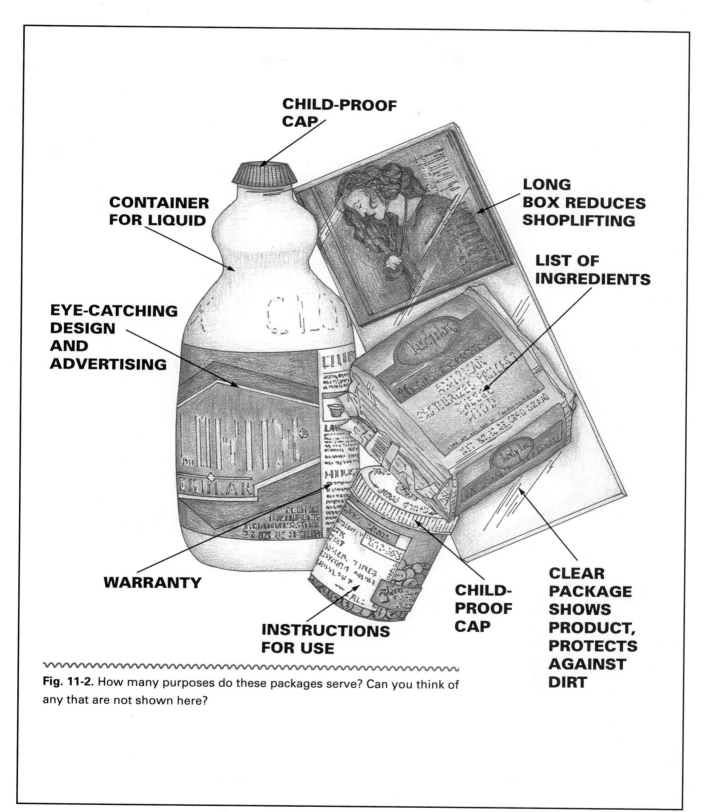

CHILD-PROOF CAP

CONTAINER FOR LIQUID

LONG BOX REDUCES SHOPLIFTING

LIST OF INGREDIENTS

EYE-CATCHING DESIGN AND ADVERTISING

WARRANTY

INSTRUCTIONS FOR USE

CHILD-PROOF CAP

CLEAR PACKAGE SHOWS PRODUCT, PROTECTS AGAINST DIRT

Fig. 11-2. How many purposes do these packages serve? Can you think of any that are not shown here?

Fig. 11-3. Warehouses are used to store large quantities of items.

Package Design

When a manufacturer decides to make a product, the specifications are sent to a packaging manufacturer. The package is then designed. The product's size, shape, and special needs are all taken into consideration. Artists design any labels or advertising.

Storage

Products are held in storage until they leave the factory. **Storage** is the holding of products until they are sent to the customer. The storage area keeps them safe from damage or theft.

Most products are stored in warehouses. A **warehouse** is usually a big building with large, open spaces. Fig. 11-3 (pp. 188-189). Some contain storage racks, bins, or trays for smaller items. Huge doors allow products to be easily moved in or out.

The size and number of products determines the size of the warehouse. A furniture manufacturer needs a large building. A maker of paper clips and other fasteners may require less space.

Stock clerks know where all the items in the warehouse are located and how many there are. Computerized storage systems may rely on pick-and-place robots to put items away or find them again.

Bar codes, like those on items in the grocery store, are now being used to keep track of products in storage. Fig. 11-4. Computers scan the bar codes and keep records of where the products are located and how many are still on hand.

► ► ► **FOR DISCUSSION** ◄ ◄ ◄

1. **What unusual packages have you seen lately? Describe one. Why did it catch your eye? Do you think it was a successful package to use? Why or why not?**

■ **Design a package for your student company's product. Be sure to think about what the package needs to do. Must it protect the product? Advertise it? Keep it together?**

Fig. 11-4. This bar code is read by a scanner linked to a computer. The computer then keeps track of how many items are on hand.

■ **Sometimes packages can be wasteful of raw materials or harmful to the environment. Collect examples of such packages and create a display. Label the packages as to why you chose them.**

Marketing

Marketing is everything a company does to sell its products. A good marketing plan is important to the financial success of a company.

As you learned in Chapter 3, researchers begin gathering information about customers before the product is made. The type of customer the manufacturer wishes to reach determines the marketing methods used. Some of these methods can be costly. The expense must be added to the cost of the product.

Advertising

As you know, some advertising is done on product packages. Other advertising is done in print, on TV or radio, and in special ways, such as on T-shirts.

Printed ads appear in newspapers, magazines, journals, billboards, and direct mailings. Newspapers and magazines reach many potential customers. Many magazines are sold all over the country. Journals are magazines read by special-interest groups. A company that makes medicines will advertise in medical journals. Makers of tractors will advertise in journals that farmers read.

Billboards are meant for quick messages. Drivers, who must also pay attention to the road, must be able to get the message in a few seconds. Fig. 11-5.

Fig. 11-5. Do you think drivers will get these messages quickly?

How TV Commercials Are Produced

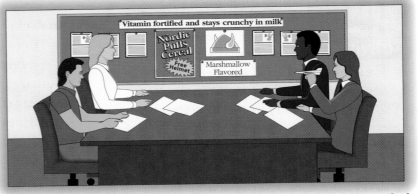

Many manufacturers hire advertising agencies to produce their commercials. They tell the agency what is wanted. In creative meetings, workers at the agency come up with several ideas.

The ideas are given to artists and writers to develop. Then more creative meetings are held. The best idea is selected. The writer writes a final script. The artist draws a storyboard. This shows in pictures how the commercial will look. The script and storyboard are shown to the manufacturer for approval.

Actors are hired and the commercial is taped. Workers in the studio's control room add music and other special effects. Air time is purchased, and the commercial then appears on TV.

Fig. 11-6. Many people are involved in the production of a commercial. The manufacturer and advertising agency work together to make sure the commercial sends the right message.

Direct mailings are ads that are sent directly to customers' homes. They are often called "junk mail." Computers enable marketing companies to "target" certain groups of people. Information such as age, income, education, occupation, and location are all fed into the computer. The computer then produces a list of those who might be interested in buying. The advertisement is then sent to them.

TV and radio ads are meant to reach the general population. TV advertising is very expensive. Depending upon how often it is run, a commercial may cost a company millions of dollars. Fig. 11-6.

Sales

Most companies have a staff of salespeople or factory representatives. The salespeople visit stores that might be interested in carrying the product. Fig. 11-7.

Salespeople need to be experts on their company's products. Often, they earn a percentage of each sale. The more they sell, the more they earn.

Fig. 11-7. These oil company representatives are taking orders for natural gas from distributors.

▶▶▶ FOR DISCUSSION ◀◀◀

1. Try to remember a commercial you saw on television last night. What was the product? Why do you think you remember the advertisement? Did it persuade you to buy the product?

_____

Company Activity

■ As a class, produce a 30-second radio or TV commercial for your student company's product. Think about why someone would be interested in buying the product. Tell customers what's good about it. Tape the commercial and play it for other classes.

Extension Activity

■ Visit several stores and observe three different salespeople at work with customers. Write a brief paragraph about each salesperson's technique. Was it successful or not? How do you think it could have been improved?

Distribution

Distribution is the moving of products from the factory to customers. Products can be carried by land, on waterways, or by air. How fast the product is needed and how far it must go determine the method used.

Air freight is costly but fast. It is used for fairly small products that are usually needed immediately. For example, medicine needed urgently might travel by air. Trains carry large and small products over long distances. Trucks are used for both long and short trips. Trains and trucks take longer than air but are much cheaper. Products that travel overseas usually go by ship. Fig. 11-8.

Wholesalers and Retailers

Some products, such as airplanes, are sent directly to the customer who will use them. Others pass through a distribution network. A distribution network links wholesalers and retailers. A **wholesaler** purchases products from manufacturers and then sells them to retailers. **Retailers** sell directly to consumers. Fig. 11-9.

Some wholesalers carry several kinds of products. Others specialize in one product. They store the products in warehouses until a retailer places an order. The items are then shipped out.

Fig. 11-8. This tower is being unloaded at a chemical plant in Belgium.

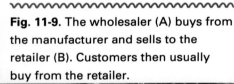

Fig. 11-9. The wholesaler (A) buys from the manufacturer and sells to the retailer (B). Customers then usually buy from the retailer.

A

B

Retailers sell products in stores or through catalogs. Because they do not have much storage space, they do not buy in large quantities. They buy just enough to fill their current needs.

A few retailers, such as furniture stores and supermarkets, usually buy goods directly from the manufacturer. The manufacturer ships the products to the retailer's warehouse. The product is then sent to the retailer's stores.

Marking Up the Price

Wholesalers have their own costs of doing business. They also want to make a profit. Their costs plus profit must be added into the price the wholesaler charges. Retailers, too, must raise the price to cover their costs and to make a profit. The difference between the purchase price and the selling price when a product changes hands is called the **markup**. Usually, a product has been marked up a number of times before it reaches the consumer.

▶▶▶ FOR DISCUSSION ◀◀◀

1. **Have you ever shopped in a store that advertised "wholesale" prices? Do you think such prices really are wholesale? Why or why not?**

Company Activity

■ **Determine what it costs to produce each item manufactured by your student company. Calculate markups of 10%, 25%, and 50%. What markup should your company use? Why?**

Extension Activity

■ **Contact your local newspaper. Ask how the paper is distributed from the printer to the different outlets. Draw a flow chart of this distribution process.**

Servicing

Most of the time, customers have no problems with a product. However, on occasion something goes wrong. **Servicing** takes place when a manufacturer supplies parts and labor to correct defects in a product. Fig. 11-10.

Servicing is usually done under a warranty. The warranty is good for a certain amount of time. It also usually states what is covered. For example, labor on a TV set may be covered for ninety days. However, parts on the TV set may be under warranty for a year. Other products may have longer warranties. Wood furniture is often warrantied for up to five years against defects.

Fig. 11-10. This appliance manufacturer provides servicing to customers.

What about items that cannot be serviced, such as cosmetics or dog food? In this case, the product is returned to the store or mailed to the manufacturer. The purchase price is usually refunded.

▶▶▶ FOR DISCUSSION ◀◀◀

1. Many years ago, manufacturers did not use warranties. Their motto was Caveat emptor, which is Latin for "Let the buyer beware." Since warranties add to the cost of products, do you think they are a good idea or not? Give your reasons.

■ **Write a warranty for the product made by your student company. Tell what kinds of defects it covers, how long it is good for, and what your company is willing to do to correct any problems.**

Recycling

In Chapter 5 you learned that many materials can be recycled. This helps reduce manufacturing costs and saves resources. Most manufacturers do a certain amount of recycling. Such things as metal scraps may be melted down and reused. Solid waste is collected and placed in storage bins. The wastes are purchased by recycling companies and used to make other products. Fig. 11-11.

Many communities have started recycling programs. People take glass bottles, metal cans, and newspapers to collection centers. The materials that make up these items can all be reclaimed for future use.

Fig. 11-11. Materials you take to a recycling center are picked up and reused to make new products.

▶▶▶ FOR DISCUSSION ◀◀◀

1. Environmental protection laws have forced many manufacturers to "clean up their acts." However, the air and water are still polluted. Should these laws be tougher on manufacturers? Give reasons for your answer.

■ **As a class, set up a recycling program for your neighborhood or school. Make posters telling people where to bring their bottles, cans, or papers. Then arrange to take these items to a recycling center.**

Chapter Highlights

● A price for a product is set by totaling all the costs and then adding the amount of profit desired.

● Packaging refers to any container that holds a product.

● Marketing is everything a company does to sell its products. Included in marketing are advertising and sales.

● Distribution is the moving of products from the factory to the customer. Wholesalers and retailers both mark up the price of a product.

● A manufacturer services a product by supplying parts and labor to correct any defects.

● Recycling helps reduce manufacturing costs and saves resources.

Test Your Knowledge

1. What is the name for a retail price set by the manufacturer?

2. What is the name for the computer codes printed on packages?

3. Give two reasons why products are packaged.

4. Where are most products stored?

5. Define marketing.

6. Name three kinds of printed ads.

7. What is the name of the process of moving products from the factory to customer?

8. From whom does a wholesaler purchase products?

9. When a product changes hands, what is the increase in price called?

10. What is the primary disadvantage of using air freight to ship products?

Correlations

LANGUAGE ARTS

1. Write "WAREHOUSE" in a vertical line. Use each letter as the beginning of a noun that names an item found in a warehouse of some kind.

SOCIAL STUDIES

1. Research the history of packaging. Ask your parents how products used to be packaged. How is milk, for example, packaged differently now than it was years ago?

SCIENCE

1. Write a report telling how the laser scanner at the grocery store checkout counter works.

MATH

1. Obtain an order form from a mail order catalog. Select 3 items you will "buy." Fill out the form. Be sure to include shipping and handling charges, sales tax, and any other applicable fees.

You·Can·

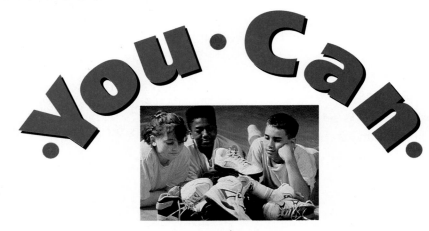

— Make a Difference —

Better, Wiser Shopping with the Z-Team

Which athletic shoes give you the most speed on the basketball court? If you wanted to invest your allowance, where would you put it?

Zillions magazine tries to supply the answers to questions like these from young people like you. *Zillions* magazine is a consumer magazine for teens and younger kids. It uses kid experts, known as the Z-Team, to test products. What the "experts" find out is then shared with *Zillions*' readers.

Z-Team members must follow strict rules. They must be fair when they are testing a product. They must be accurate. They try not to let themselves be influenced by TV commercials, advertisements, or their friends.

For example, when the Z-Team tested athletic shoes, they let their feet do the "thinking." Three brands of shoes were tested. Testers walked, ran, and jumped in all three. As their times were recorded by a stopwatch, they sprinted across a gym floor. The results of the testers' findings were published in *Zillions*.

Many young people are concerned about the environment. As a result, Z-Team members checked out recycling in fast-food restaurants. They analyzed the packaging used there. They found a lot of unneeded packages. Of the 24 fast-food places the Z-Team visited, only one had recycling bins.

Zillions is published by Consumers Union, Mt. Vernon, NY.

Sometimes advertisers take advantage of kids. They design ads that look like part of the content of the magazine. Younger kids are often fooled. They think that what the ads say is all true. *Zillions* does not use such advertisements. In fact, a regular feature called "The Sneaky Sell" helps make young people aware of misleading ads.

Another regular feature of *Zillions* is "Money Talk." It is for those who want to invest money wisely. As an experiment, the magazine has invested $500. In each issue a report is given as to how the investment is doing.

Zillions also covers other issues that are important to younger readers. Sometimes Z-Team members give advice. For example, what do you do if a friend borrows money and

does not pay it back? One member suggested, "Be honest and tell the friend how you feel."

How does someone become a member of the Z-Team? First, you fill out a survey. The survey asks you questions about how you spend your money. Then you are also asked to rank products that you use everyday. One hundred lucky kids are then chosen to become official testers.

CONSTRUCTION

Activity Brief:
Creating a Shelter for the Homeless

PART 1: Here's the Situation...........

The construction industry started thousands of years ago when people first began to build shelters. They may have used stones, mud, or branches and leaves from trees. As they developed better skills and materials, the construction industry changed. However, the human needs it serves are the same today as they were in times past.

Construction technology is our use of tools, materials, and processes to create structures. When we think of construction, we usually think of buildings. However, construction also includes such things as highways, dams, towers, and pipelines.

In this section you will learn about the types of construction and the materials, tools, and processes used. You will also discover how construction is planned for and why quality is important.

PART **2**: Your Challenge..........

The need for shelter is a very important one. Shelters protect people from heat or cold and from wind, rain, snow, and too much sun. They also keep people safe from insects, wild animals, and other dangers. People could not live very long without shelter, except in a few places where the climate is always mild.

People can be without shelter for several reasons. They may be too poor to obtain it. This is true of some of the people who sleep in parks and other public places and who have no homes. During natural disasters, such as earthquakes and floods, people's regular homes may be destroyed. Wars are another cause of homelessness. During the war between the United States and Iraq, thousands of Kurds were driven from their homes.

In all of these cases, people must find other places to stay. Temporary shelters have to be created.

For this activity, you will design a temporary shelter for homeless people. Ideally, it will be something that could be sold to rescue agencies, such as the Red Cross, and used in many parts of the world. You will make working drawings and a floor plan. Then you will create a scale model of your structure.

PART **3**: Specifications and Limits..........

Your shelter will need to meet certain standards. Read the following specifications and limits carefully before you begin.

1. Your shelter must be:
- *temporary* — easy to put up or take down
- easy to transport in large numbers by plane or truck

Design Tip:
How will you anchor your shelter to the ground? The wind must not be able to blow it over. Rain and snow must not be able to get in underneath.

FLOOR PLAN

34'

ALL AISLES

TABLE

INSTR.

CABINETS

- made of fairly inexpensive materials
- equipped with a door and two windows
- of a size to sleep a family of four people (approximately 140 square feet)
- usable in a variety of climates and able to withstand heat, cold, wind, rain, and snow
- equipped with heat (as needed), fresh water, and sanitation facilities
- easily maintained and repaired

2. You must hand in the following:
    - a log sheet of your work
    - a written needs assessment of the project
    - at least 3 sketches of designs you considered
    - a finished working drawing of the shelter, showing front, back, and sides and including dimensions
    - a floor plan
    - a model of the shelter, built to a scale of ¼" = 1'
    - instructions for assembly
    - a paragraph describing the shelter and telling why you chose this design and materials
    - A list of any resources you used (books, magazines, museum displays, interviews with people, etc.)

3. You will be evaluated on the basis of:
    - originality
    - how usable your design is
    - how well you have followed design specifications
    - how well you took different conditions into account
    - how well you researched your design

PART 4: Materials..........

There are many materials you might use to create your design. Here is a list of possibilities.

Materials

paper and pencils
drafting tools and equipment
computer and CAD software
cardboard, plastic, wood, sheet metal, foam core board, rubber bands, dowels, fasteners, and other materials generally available in the lab

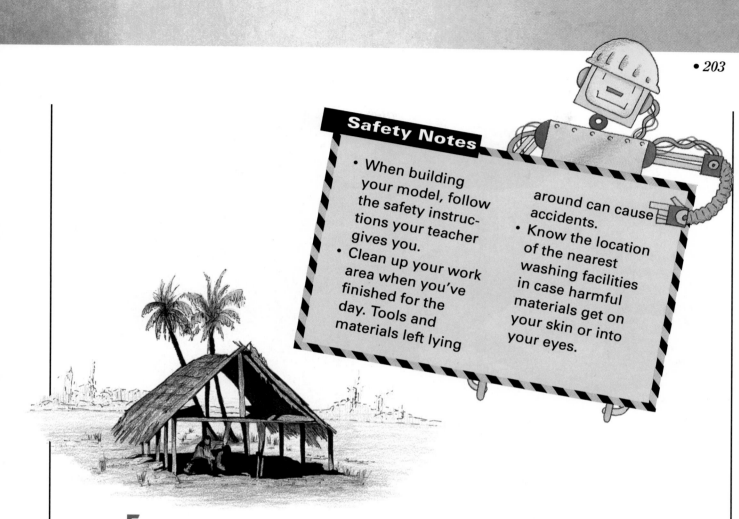

PART 5: Procedures

The design you choose and how you build it will be up to you. Still, there are certain steps to follow that will make your work easier.

1. Be sure to keep a daily log of your work.

2. Read all the information provided with this activity. Review the steps in problem solving.

3. Using your school or community library, do some research. For example, you might look up shelter in an encyclopedia. You might also research temporary shelters currently being used. Have you ever lived in a temporary shelter? If so, think back over your own experience.

4. Make some sketches of possible designs. Choose the one you like best and submit it to your teacher for approval.

5. Make the finished drawings.

6. Build the model of your design.

7. Finish the written portion of the activity.

PART 6: For Additional Help...........

Information about the following is found on these pages in the text:

architectural working drawings, pp. 62-65
floor plans, p. 62
hand tools, pp. 114-117
insulation, pp. 250-251
materials, pp. 94-108
planning for construction, pp. 227-234
power tools, pp. 118-123
problem solving, steps in, pp. 24-26

You might also want to research these related subjects in reference books:

the Red Cross and other disaster relief organizations
structures made of foam and other unusual materials
collapsible structures
tents

Consider also magazines, such as *Popular Science*.

PART 7: How Well Did You Meet the Challenge?...........

When you've finished your design, evaluate it. Does it do what it is supposed to do? Ask yourself the following questions. Your teacher and classmates may take part in this evaluation.

1. Is your design of adequate size? Does it have water, heat, and sanitation facilities?

2. Will your design withstand heat, cold, rain, wind, and snow?

3. Can it be easily shipped, assembled, and taken down?

4. If your own home were destroyed for some reason, would you want to live in the shelter you designed?

5. How might you improve your design?

PART **8**: Extending Your Experience..........

Think about the following questions and discuss them in class.

1. Natural disasters, such as hurricanes, take many lives each year. Suppose that human beings could stop such natural disasters before they happened. What impacts might occur as a result?

2. Disease often spreads rapidly among disaster victims in refugee camps. Why do you think this is so? What could be done to prevent it?

3. Read on page 276 the story of teens who built a shelter for the homeless in the South Bronx, New York. Would such a project benefit your own community? Whom could you contact to learn more about building one?

Did you know...?

- In Japan, where earthquakes are frequent, homes are made of light materials, which are safer. Roofs are of light bamboo sticks. Partitions are made of paper.
- In hot, dry climates, walls are thick to keep out the heat. Windows are narrow and there are few of them. Because there is little rain, roofs are flat.
- Steep, pointed roofs allow rain to run off easily. Bundles of grass or thatch laid on the roof also prevent rain from getting inside.
- In Tibet, many people live in tent-like "yurts." A yurt is made by covering a framework of three branches with felt.
- The ancient Romans were the first to develop central heating. Rows of earthenware pipes were laid under the floor. Hot air or water was run through them.

CHAPTER 12

Getting to Know Construction

Introduction ....

You have already learned about the ways in which manufacturing systems operate. Construction systems work in similar ways. They require planning and quality control. There are many different processes used. The end result is a product that satisfies human needs and wants.

In this chapter you will learn about the different types of construction. You will learn about buildings, civil construction projects, towers, pipelines, and other special structures.

After studying this chapter, you should be able to ....

Discuss the different types of construction projects.

Name the three types of buildings.

Describe the different kinds of civil construction projects.

Briefly discuss towers, pipelines, and other structures.

Words you will need ....

cantilever

civil construction
  projects

cofferdams

commercial buildings

heavy construction

industrial buildings

light construction

locks

navigation canals

residential buildings

truss

Buildings

Buildings come in many shapes and sizes. Some, like single-family houses, garages, and garden sheds, are fairly small. The materials and methods used to make these structures are referred to as **light construction.**

Factories, office buildings, and high-rise apartments require different materials and methods. For instance, a steel or concrete frame is usually used. The frame is attached to the foundation and supports the weight of the structure. The methods and materials used to build these structures are referred to as **heavy construction.** Heavy construction methods and materials are also used to build roads, bridges, airports, and tunnels.

Buildings may be classified as to their purpose. Their purpose may be residential, industrial, or commercial.

Residential Buildings

Residential buildings are those in which people live. Single-family houses are designed for just one family. Multiple-family buildings house more than one family. They include apartments and condominiums.

The frames for most single-family houses are made of many pieces of wood. Fig. 12-1. The wood and other materials are brought to the building site and assembled there.

Prefabricated houses are built at a factory in sections. The sections are then brought to the site, where they are assembled. Manufactured houses are built entirely in a factory. They are then transported to the site.

Industrial Buildings

Factories and oil refineries are examples of **industrial buildings.** These buildings are used by production industries. Some service industries, such as waste recycling, also need industrial buildings. Many factories require only one large building. Others may need a dozen buildings in one location. Fig. 12-2. Each building is used for a different step in the production process.

Fig. 12-1. Wood-framed buildings are a type of light construction.

Fig. 12-2. This large factory complex is part of a paper manufacturing company.

The buildings are carefully laid out so that the work of the business can be done effectively. Some buildings, such as shopping malls and restaurants, should also be attractive. This helps to draw customers.

Commercial buildings must be built according to government regulations. These regulations specify construction standards. For example, most commercial buildings must have steel frames. Others may be required to have fire sprinkler systems.

Fig. 12-3. Large public buildings, such as this city hall in Boise, Idaho, are considered commercial buildings.

Commercial Buildings

Commercial buildings are usually used by businesses, such as supermarkets and insurance companies. Libraries, schools, and churches require similar buildings even though they are not used for business. Fig. 12-3.

▶▶▶ FOR DISCUSSION ◀◀◀

1. Why do you think a single-family house and a high-rise apartment building would require different construction methods?
2. Why do you think commercial buildings must be built according to government regulations?

Extension Activity

■ Take a walk through the neighborhood around your home or school. Make a list of the different types of buildings you see. Which construction method was used in each case, light or heavy?

Civil Construction Projects

Civil construction projects are those that serve the public in some way. They include highways, bridges, tunnels, airports, dams, and canals. These projects are important for such things as public transportation and generating electricity. Earthmoving equipment and other heavy machinery is usually required during construction.

Fig. 12-4. To make the pyramids at Gizeh in Egypt, workers drew enormous stone blocks over long distances.

IMPACT

Politics can have an impact on construction. For example, a county or state can lower taxes and create other advantages. Then factories and other companies may be encouraged to locate there. As a result, new buildings and homes for workers may be built.

Roads and Highways

Roads have been used by people throughout history. About 5,000 years ago, the Egyptians built roads for construction purposes. The roads were used to transport giant stone blocks and other materials to the pyramids. Fig. 12-4. In Europe, some of the roads originally built by the Romans centuries ago have been turned into modern highways. In the United States, the interstate highway system we now use was first laid out in the 1700s.

Fig. 12-5. Roadways require several layers. Many workers and machines are used.

Today, roads are needed for automobile transportation. Cars are an important part of daily life for most of us. Roads are also used by trucks to bring raw materials to factories, building sites, and consumers.

Building a modern highway is complex and expensive. A four-lane highway may cost several million dollars *per mile*. Because highways affect the environment, an impact study may have to be done. Maps and surveys may be made to select the best route. Sometimes aerial photographs are necessary. Because roads must have a good foundation, highway engineers must test the soil.

Before construction begins, the soil underneath must be packed down and prepared for paving. Then a layer, called a base, is added. Some roads may have several base layers. Finally, the road surface is laid on top. Fig. 12-5.

Bridges

Bridges are used for carrying people or vehicles over something, such as a river or railroad track. The first bridges were probably trees that fell across streams. Later, people made their own bridges by cutting logs with stone tools. In Europe during the Middle Ages, wooden drawbridges were used to protect castles. The bridge could be raised or lowered across a pool of water, known as a moat, that circled the castle. In the 1700s, iron was popular as a bridge material. Today, we use steel and concrete.

Bridges must be strong. They must support their own weight and the weight of people and vehicles crossing them. They must hold up under rain, snow, high winds and changes in temperature. If they are over rivers, they must be able to resist the force of moving water. If they are in areas where earthquakes are common, they must resist shaking. Fig. 12-6.

If not cared for, bridges can become damaged or even collapse. For this reason, they are maintained and inspected regularly. Regular painting helps reduce the corrosion of steel parts. Inspections help make sure there is no damage from such things as floodwaters.

Fig. 12-6. This overpass bridge collapsed during an earthquake in San Francisco a few years ago.

Bridge Design

Many things must be considered when designing a bridge. How long must it be? How much weight will it carry? What sort of obstacle will it cross? Every bridge has its own special requirements for engineers to consider. Two types of bridges are shown in Fig. 12-7.

Fig. 12-7. How do the supports for these bridges differ?

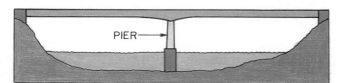

BEAM BRIDGE. This type of bridge uses steel or concrete girders. These girders, or beams, are horizontal supports on which a concrete roadway rests. When supported by a number of piers, or vertical supports, this simple bridge can cover long distances.

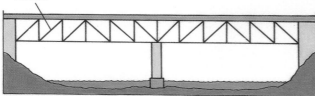

TRUSS BRIDGE. A truss is a triangular framework that can carry loads. It can be above or below the roadway. A truss bridge is strong and economical to build. Trusses may be used as part of other bridge designs to add strength.

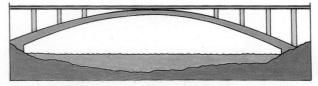

ARCH BRIDGE. This kind of bridge uses an arch to carry the load. The arch is supported on each side by a type of support called an abutment. The roadway can be above the arch or below it.

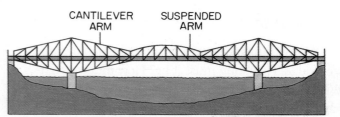

CANTILEVER BRIDGE. A cantilever is a beam that extends from each end of the bridge. Cantilevers do not reach all the way across. They are connected in the middle by a part called the suspended span.

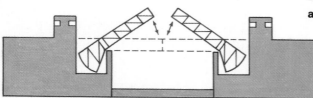

MOVABLE BRIDGE. Some bridges, such as the bascule bridge shown here, can be opened to allow large ships to pass. Bascule bridges open by tilting upward. Lift bridges have a section of roadway that moves between two towers. Swing bridges have a section that swings sideways.

Fig. 12-8. These are the seven basic bridge designs.

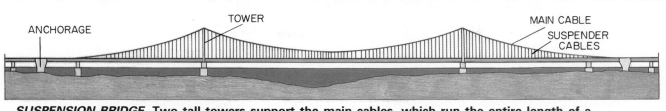

SUSPENSION BRIDGE. Two tall towers support the main cables, which run the entire length of a suspension bridge. The cables are secured in concrete at each end. Other cables are attached to the main cables. These hanging or suspended cables are used to support the roadway. Sometimes a second road can be added below the first. Suspension bridges are used for long distances. The Verrazano-Narrows Bridge in New York is over two and one-half miles long.

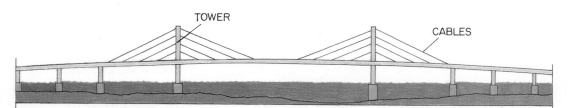

CABLE-STAYED BRIDGE. Cable-stayed bridges are like suspension bridges except that the cables are connected directly to the roadway. Most cable-stayed bridges have been built outside the United States. American engineers are not sure that these bridges are strong and durable enough for use in this country.

Engineers must have a thorough understanding of forces and loads in order to design a bridge. Features such as arches, trusses, cantilevers, and cables all help the bridge carry weight. There are seven basic bridge designs. These are shown in Fig. 12-8. Sometimes two or more designs are combined in one bridge.

Dams

Dams control or stop the movement of water in a river or stream. A dam can control flooding. It can also be used to generate electricity (see Chapter 4). The lake that forms behind the dam can be used to supply water for drinking and irrigation. It can also be used for recreation. Fig. 12-9.

Fig. 12-9. Lake Quachita, formed behind a dam near Hot Springs, Arkansas, is used for recreation.

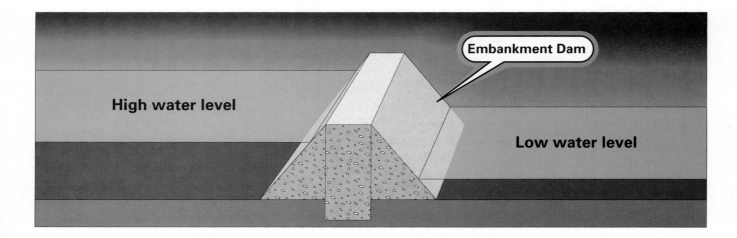

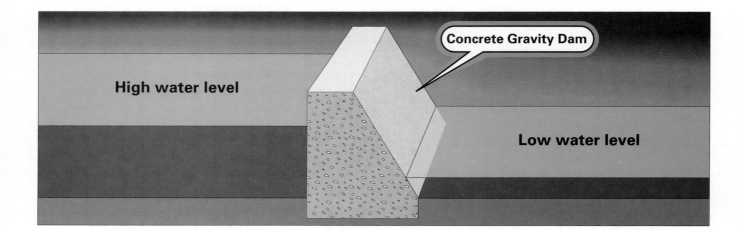

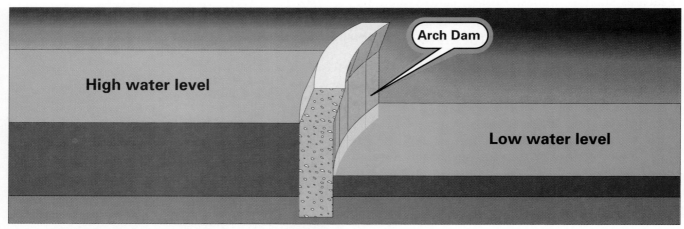

Fig. 12-10. An embankment dam uses the weight of the earth to hold back the water. Concrete gravity dams are very expensive and usually run across wide valleys. The arch dam is very strong. The curve of the arch is what holds back the water.

Fig. 12-11. Tunnels go through or under an obstacle.

Dams can be made of many different materials. These include concrete, rock, earth, stone, and wood. The design of the dam and the materials used depend on local conditions and the dam's purpose. Fig. 12-10. Dams can take ten or more years to complete.

First, geologists study the site. Geologists are scientists who study rock and earth formations. They learn if the rock at the site is suitable for a foundation. In addition, they study the river or stream to determine how large a lake will be formed. The impact on the environment must also be considered.

During construction, temporary dams, called **cofferdams**, are made first. The cofferdams shift the flow of water to clear the site. Then the permanent dam can be built. Tunnels may also be used for this purpose.

Tunnels

Bridges go over obstacles, but tunnels go under or through them. A tunnel can take highways or railways through mountains or under cities and rivers. Fig. 12-11.

Earth tunnels are dug in soil or sand. Because soil and sand can collapse easily, earth tunnels are dangerous to build. As they are dug, concrete sections are usually installed for support.

Tunnels through rock are created by blasting or by using giant boring machines. Fig. 12-12. For tunnels under water, pre-made sections are floated to the site, sunk, and connected.

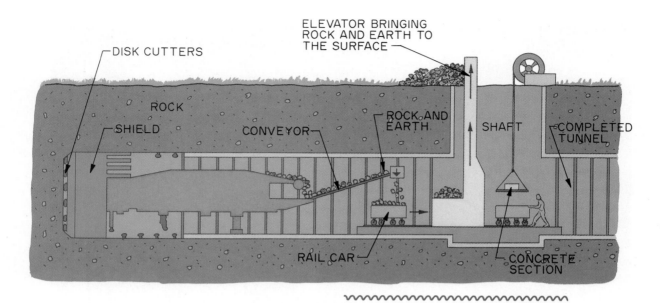

DISK CUTTERS

ELEVATOR BRINGING
ROCK AND EARTH TO
THE SURFACE

ROCK

SHIELD

CONVEYOR

ROCK AND
EARTH

SHAFT

COMPLETED
TUNNEL

RAIL CAR

CONCRETE
SECTION

Fig. 12-12. Giant boring machines can cut tunnels through soft rock.

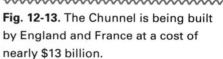

Fig. 12-13. The Chunnel is being built by England and France at a cost of nearly $13 billion.

One giant underwater project now being built is the Chunnel. The Chunnel is a 31-mile long tunnel beneath the English Channel. Fig. 12-13. When it is finished, it will connect England and France. Trains will travel through it. The Chunnel is the largest civil engineering project in the history of Western Europe.

Airports

The first airports were little more than cow pastures. They worked well for the first planes, which were light. Then planes got heavier and flew in all kinds of weather. Runways with hard surfaces became necessary. The first city-owned airport was built in Tucson, Arizona, in 1919. Fig. 12-14.

Today, roads, parking lots, a passenger terminal, an air-traffic control tower, and cargo, catering, rescue, and maintenance facilities are all needed. Runways and taxiways for planes must be built to strict standards. Fig. 12-15. Some newer airports include ground transportation systems, such as monorails. As a result, many people work together to build an airport. They include architects, engineers, aviation experts, and government officials.

Fig. 12-15. Many layers of material are needed to make a runway. It must support heavy planes traveling at high speeds.

Fig. 12-14. Tucson was the first American city to build its own airport. This photo shows how it looked in 1927. How was it different from an airport of today?

IMPACT

Not all impacts of construction are positive. Problems created can be temporary. The closing of one lane of a highway while it is being widened is one example. Others may be permanent. The expansion of an airport may encourage airlines to add flights. This in turn increases the noise problems that already exist.

Canals

Canals are long trenches filled with water. They may be used for irrigation by carrying water from one place to another. When canals are used for transportation, they are called **navigation canals**. Canals have been used for transportation for more than 4,000 years.

Navigation canals connect two bodies of water. For example, the Panama Canal is located in Central America. It connects the Atlantic and Pacific Oceans. A ship sailing from New York City to San Francisco, California, saves 7,800 miles by using the 51-mile long Canal. Without it, the ship would have to sail all the way around South America.

Canals are planned by engineers who study the land and select the best route. A lot of digging, using heavy equipment, must be done. Environmental impact studies are made when considering canal sites.

Sometimes locks are needed. **Locks** are special chambers built into the canal that allow a ship to be raised or lowered on its way through. The locks are needed when land or water levels change along the canal route. Fig. 12-16.

Fig. 12-16. Locks raise or lower ships as they pass through a canal.

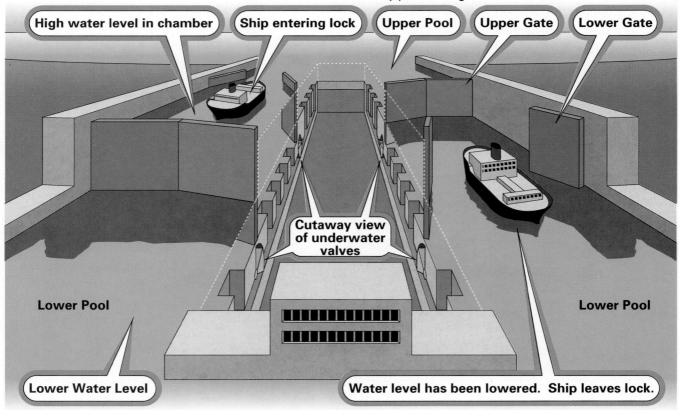

High water level in chamber · Ship entering lock · Upper Pool · Upper Gate · Lower Gate

Cutaway view of underwater valves

Lower Pool

Lower Pool

Lower Water Level

Water level has been lowered. Ship leaves lock.

▶▶▶ FOR DISCUSSION ◀◀◀

1. What impacts might occur when a major highway that goes through the center of a town is replaced by an alternate that goes around the town?

2. What factors must be considered when selecting the type of bridge to build over a certain river? What factors must be considered when deciding where to locate the bridge?

3. What concerns would you and your family have if a new airport was to be built in your town?

4. Few canals are being built today. Why do you think this is so?

_____

Towers

Towers are tall, narrow structures. They have many uses. They may be tall buildings, commonly called skyscrapers, that are used for offices. They may be built as memorials. The Washington Monument is a tower built to honor George Washington. The Eiffel Tower in France was part of an exhibition for the public in 1889. The world's tallest freestanding structure is the Canadian National Tower in Toronto, Canada. It soars to a height of 1,815 feet. Fig. 12-17.

Many communities have water towers. The tower helps create the pressure needed to deliver water to the people who live nearby. Electrical towers support the cables that carry electric current. Television and radio towers hold the antennas that broadcast TV and radio signals.

Extension Activity

■ 1. Make a model of one of the bridges shown in Fig. 12-8. Use materials such as cardboard, balsa wood, clothespins, straight pins, and glue.

■ 2. Research the construction of one of the following projects and report on it to the class. Include information on how long the project took, how many workers were needed, the purpose of the project, and its location.
• The Erie Canal
• The Pan American Highway
• Hoover Dam
• The Holland Tunnel
• The Brooklyn Bridge

▶▶▶ FOR DISCUSSION ◀◀◀

1. Some skyscrapers are so tall that you must take not just one elevator, but a series, to reach the top. Would you enjoy living or working in such a building? Why or why not? What are some advantages and disadvantages of skyscrapers?

_____

Company Activity

■ 1. If your student company is building a structure, decide what that structure will be. If it is to be permanent, select a site.

Fig. 12-17. The Canadian National Tower is a famous landmark in Canada.

Pipelines

Pipelines are used to carry liquids or gases over distances. Pumping stations along the pipeline maintain pressure inside. The pressure helps move the material through the pipe. Crude oil, refined petroleum, and natural gas are all transported through pipelines.

When pipelines are buried, huge trenches are dug along the route. Then the pipes are laid and the sections welded together. X-rays and other tests are made of the welds to be sure they are tight. Then the pipe is covered with earth. Fig. 12-18.

Fig. 12-18. These pipes are part of the Trans-Alaska pipeline and are usually snow-covered.

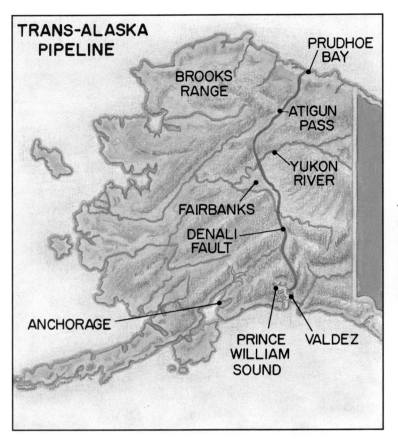

Fig. 12-19. The Trans-Alaska pipeline, stretching for 800 miles, brings oil from Alaska's northern reaches to Valdez on the southern coast.

Twenty-five percent of the crude oil used in the United States is carried by the Trans-Alaska Pipeline. It stretches for 800 miles from Prudhoe Bay in North Alaska to Valdez on the south coast. Fig. 12-19. Rough land and bad weather made building the pipeline difficult. Some parts are above ground and supported by bridges. Other parts are buried. Corrosion has been a serious problem. The pipeline must be constantly maintained. This is quite expensive.

The pipeline has benefited Alaska's economy. Both its construction and continued operation have brought in much money to the state. However, environmentalists are worried that the pipe may leak and cause damage to soil and water.

▶ ▶ ▶ **FOR DISCUSSION** ◀ ◀ ◀

1. Some pipelines stretch across borders between one state or one country and another. What problems do you think this could cause?

_____

Extension

Activity

■ **Research the major pipelines of natural gas across the United States. Draw their locations on a map.**

Other Structures

Sometimes structures are built for special reasons. They do not fit into the ordinary categories. Structures built in outer space are one example. One day NASA hopes to build a space station that will be used for research. Parts of the station will be built on earth and then carried into space on the Space Shuttle.

Off-shore oil rigs are another example. The huge rigs are located on the ocean, many miles from land. Fig. 12-20. People live and work on the rigs while they look for oil. The rigs are also used to pump crude oil when it is found.

Some structures are built just for recreation. Walt Disney World is an example. Fig. 12-21 (p. 224). Its buildings, like the Walt Disney World Dolphin Hotel, are specially designed for entertainment. Golf courses, arenas, and stadiums are other examples.

Toronto's Sky Dome is the first stadium to have a fully retractable roof. This means the roof can be pulled back. Baseball fans can watch a game indoors or outdoors, depending on whether the roof is open or not. The stadium also houses an eleven-story hotel and several restaurants. The Sky Dome complex is near the Canadian National Tower and is shown in Fig. 12-17, on page 220.

▶▶▶ FOR DISCUSSION ◀◀◀

1. To save energy, some people have built underground homes. Do you think you would like to live or work in an underground building? Give your reasons.
2. What difficulties do you think workers will encounter when building the space station?

Extension Activity

■ Suppose your school is planning to place a dome over the football field to protect the players and fans from the weather. Design and draw a sketch of the dome. List on a separate sheet of paper all the things that must be considered before the dome is actually built.

Fig. 12-20. Helicopters and boats bring everyday necessities to the workers who live on oil rigs.

Fig. 12-21. The design of the Dolphin Hotel at Walt Disney World in Lake Buena Vista, Florida, is just as fantastic as the many rides and shows. Disney aspires to build a complete town one day.

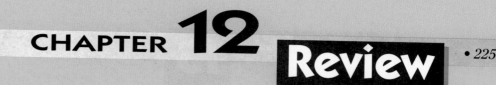

Chapter Highlights

● Light construction is used for small structures, such as single-family houses. Heavy construction is used for large structures, such as factories and office buildings.

● Residential buildings are those in which people live. Industrial buildings are used by production and some service industries. Commercial buildings are used by businesses.

● Civil construction projects serve the public in some way. They include highways, bridges, dams, tunnels, airports, and canals. Towers are tall, narrow structures. Pipelines, like the Trans-Alaska Pipeline, are used to carry liquids or gases over distances.

Test Your Knowledge

1. What kind of construction is used for houses and garages?

2. Name three structures built using heavy construction.

3. Name the three classifications for buildings.

4. Name three different types of civil construction projects.

5. List four different kinds of bridges.

6. What is a temporary dam called?

7. Why are tunnels dangerous to build?

8. What type of bridge is used for long distances?

9. What is the purpose of a pipeline?

10. What are the chambers called that raise or lower a ship as it passes through a canal?

Correlations

LANGUAGE ARTS

1. Make a crossword puzzle including types of buildings, civil construction projects, towers, pipelines, and famous examples of each. Use their definitions or descriptions as clues.

SOCIAL STUDIES

1. Write a report comparing road building techniques used during ancient times with those used today. Were there any advantages to the ancient techniques? If so, what were they?

SCIENCE

1. Study the effect of water pressure. Punch one small hole in the side of a two-liter plastic bottle near the bottom and another half-way up. Compare the distance the water squirts out from each hole. Where is the pressure greatest?

MATH

1. Construction designs include all types of geometric shapes. Make a list of all the shapes that you see in your school, a bridge, and an office building. Which shape do you see most frequently?

CHAPTER 13

Preparing for Construction

Introduction.....................................

You learned in Chapter 3 how research and planning are done for a construction project. After the project is approved, a number of tasks must be completed before actual construction can begin. The job must be organized. The field office must be set up at the site. Then the site itself must be prepared.

After studying this chapter, you should be able to.....................

Discuss how a construction job is organized.

Tell the duties of the project engineer and the construction boss.

Explain why safety is important on a construction site.

Explain why surveying is done.

Describe batter boards and tell their purpose.

Words you will need ....

batter boards
construction boss
dead load
field engineer

general contractor
live load
subcontractor
surveying

Fig. 13-1. A subcontractor, like this plumber, specializes in one type of work.

Organizing the Job

Once a construction company receives a contract, it begins to prepare for the job. A project manager organizes the money, equipment, and workers. This manager also hires subcontractors, makes the schedule, and orders materials.

Hiring Subcontractors

A **general contractor** is a company that agrees to manage the entire project. The general contractor is responsible for all the work that is done from beginning to end. Many general contractors specialize in a particular type of construction. For example, one company may construct commercial buildings. Another may construct highways.

A general contractor usually has a crew of workers. For most jobs, however, the general contractor will also hire subcontractors. A **subcontractor** specializes in one type of work, such as plumbing or painting. Fig. 13-1.

Making Schedules

All the tasks that must be done on the job are listed on schedules. Usually the job must be finished by a date stated in the contract. Some flexibility is allowed for bad weather and other problems. Schedules help the project manager keep track of what must be done and when.

Two types of schedules are shown in Fig. 13-2. The bar chart (Part A) is easy to make and understand. It shows the start and finish date for each major task. With special software, it can be drawn using a computer. The critical path chart (Part B) shows the order in which things must happen. For instance, a roof must be framed and covered before shingles are applied.

Ordering Materials

Delivery of materials must be coordinated with the work being done. Materials that arrive too early may be damaged or stolen. If they are late, work may have to be stopped. The project manager makes sure materials arrive at the proper time.

Project:_____

Date _____ **By:**_____

FORD CONSTRUCTION
CONSTRUCTION SCHEDULE

	July	Aug.	Sept.	Oct.	Nov.	Dec.	Jan.	Feb.	Mar.	Apr.
Excavation and grading	██	█								
Foundations – formwork – walls		█████████								
Rebar		██████								
Concrete		█████████								
Suspended slabs			███ █							
Slab on grade			██████							
Plumbing underground	█████									
Plumbing above ground			██████████████████							
Electrical underground	█████									
Electrical above ground		████████████████████████████████								
Mechanical – air conditioning				██████						
Roofing				███████						
Lath and plaster			██████████							
Millwork – doors – windows						████████████████				
Painting							████████			
Paving and landscaping – parking						████████				
Hardware										█
Final inspection – pick up comp.										

(A)

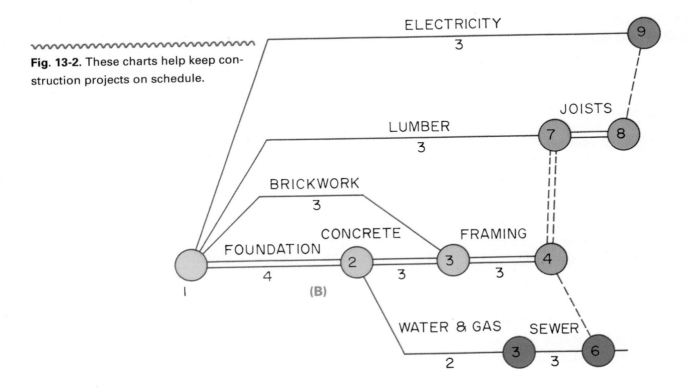

Fig. 13-2. These charts help keep construction projects on schedule.

(B)

Fig. 13-3. Paperwork and other business done at the site takes place in the field office.

Fig. 13-4. Advances in engineering have made highrises possible. Steel is used in the framework. Deeper foundations are sunk. The concrete extends all the way down to bedrock.

The cost of materials should not be greater than the figure given on the estimate. If it is, the company could lose money. The project manager also checks to make sure the costs are correct.

Establishing a Field Office

Every construction site needs a center of operations. This center is usually the field office. A small building may be built or a trailer moved to the site. Fig. 13-3. In the field office a clerk usually takes care of general business, such as payroll records.

The Field Engineer

The field office may also be used as headquarters for the field engineer. The **field engineer** oversees large projects, such as highways or skyscrapers, to be sure they are built properly. The foundation and the framework of the walls and roof are the most important parts of the structure. These must be made correctly or they will not be able to carry the expected load. Fig. 13-4.

Fig. 13-5. This construction boss keeps in touch with what is happening on the upper stories of a highrise by means of a two-way radio.

The load is the weight the structure must carry. It also includes any forces the structure will have to resist. The **dead load** is the weight of the structure itself. All the concrete, lumber, nails, and paint used to make it are a part of the dead load. The **live load** is the weight applied to the structure later on. This load will often change. It includes people who may enter and leave, furniture, and even snow on the roof.

The Construction Boss

The field office is also used by the **construction boss** or superintendent. Fig. 13-5. This person is in charge of everything that goes on at the site. The boss keeps the job on schedule and sees that the work is done properly. He or she must be a good organizer.

Different workers, such as welders, plumbers, and painters may have their own supervisors. These supervisors report to the construction boss. They receive instructions directly from him or her.

▶▶▶ FOR DISCUSSION ◀◀◀

1. Have you ever used a schedule to help you organize your work? Tell what happened. Was it successful?

2. If a live load is not permanent, why do you think it has to be considered when building a structure?

Company Activity

■ **Prepare a schedule for your student company project. Be sure all tasks are included.**

Preparing the Site

Before construction can actually begin, the site must be prepared. This includes taking safety precautions, adding roads or utilities, surveying, clearing the land, and laying out the site.

Planning for Safety

As you learned in Chapter 2, safety on the job is important both to workers and to the company. This is especially true on a construction site, which can be a dangerous place. People can fall into deep

Fig. 13-6. How many dangers can you find at this site?

holes dug for foundations. They may also trip over building materials left lying around. Objects may fall from scaffolding, platforms, or upper stories of buildings. Fig. 13-6.

Many construction sites have hard hat areas. In these areas, people must wear hard hats to protect their heads from falling objects. Safety glasses and gloves may also be required for some jobs. Railings around scaffolds or open framing protect workers from falls. Fig. 13-7.

Fig. 13-7. Steelworkers must often walk out on narrow girders. Under such conditions, little protection is possible.

Fig. 13-8. This wooden walkway protects the public from falling objects and other dangers during construction.

By law, building contractors are also responsible for the safety of the general public. A fence is often built around the site to keep curious people at a distance. Fig. 13-8. A tunnel or some sort of overhead covering may be used to protect those who pass by from falling items. During highway construction, a flagperson may redirect traffic, or an existing road may be closed.

IMPACT

A trend identified by the 1990 census was the movement of people from cities and farms to the suburbs. This shift in population has created the need for new construction in suburban areas. Taxes in these areas have gone up. Demands for public services have also increased.

Adding Roads, Buildings, and Utilities

If a new construction site is undeveloped, it will require special preparation. If there is no road leading to the site, a temporary one will have to be made. When construction is completed, the road can be removed.

A trailer or building must be located on the site to serve as the field office. Some equipment and materials must be protected from the weather and theft. Other temporary buildings or trailers may be used for this purpose. Portable restrooms are also often required.

Water for drinking and washing and electricity for lights and equipment must be installed. The field office must have a telephone. Waste containers must be provided for disposal of waste products. Until these facilities are in place, workers cannot be brought to the site.

Surveying the Site

The plans for a project tell workers what needs to be done to the site before construction begins. They also indicate how the site should look when finished.

Surveying is done to determine the exact boundaries of the property. Any mistakes can mean that the property lines are wrong. Surveyors use an instrument called a transit to measure angles and elevations and line up objects. Fig. 13-9. When the surveyors are finished, they drive stakes into each corner of the property to mark its boundaries.

Clearing the Land

The site plan is also used as a guide to clear the land. Trees and any buildings that must be protected are marked. Other trees or buildings will be removed. Bulldozers are usually used to push stumps, large rocks, and other unwanted materials out of the way. In the past, all trees and shrubs were usually cleared from a site. Today, people are concerned about the environment. As a result, natural features are often preserved.

Fig. 13-10. Before new construction can begin, the site must be cleared. This wrecker is knocking down a building.

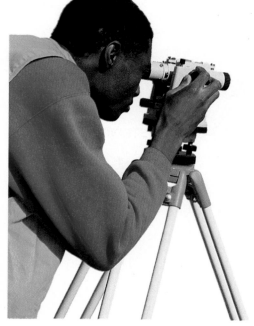

Fig. 13-9. This surveyor is using a transit to determine boundaries of a piece of property.

Sometimes existing structures must be torn down. Bulldozers or wrecking balls may be used. Fig. 13-10. In the case of very large structures, explosives may be required. When this is done, care must be taken not to damage the surroundings. Companies that specialize in this kind of work are usually hired for big wrecking jobs.

Laying Out the Site

The site plan shows how far the structure will be from the edges of the property. Using the site plan as a guide, the builder marks the exact location of the structure. The stakes left by the surveyors are used as a guide.

New stakes are driven into the ground to mark the structure's corners. **Batter boards** are placed four to five feet from each corner. These are boards held in a horizontal position by another set of stakes. String is placed over these stakes at each corner. Then the strings are run between the batter boards. The strings cross where the corner stakes are located. The strings outline the boundaries of the structure. Fig. 13-11. The corner stakes can then be removed. The strings and batter boards remain and are used as a guide for excavation.

▶▶▶ FOR DISCUSSION ◀◀◀

1. How do you think a surveyor lines up the boundaries of a site?

2. Name as many possible hazards at a construction site as you can think of. Tell why you think each is dangerous.

3. How do you think using explosives to clear a site could be dangerous to the environment?

1. Conduct a safety inspection of the site for your student project. What precautions are necessary? Are safeguards needed for the public? Make a list of the requirements. Then appoint a safety supervisor to be sure all the requirements are met during construction.

2. Using scrap lumber, create a set of batter boards. Using the batter boards and string, lay out the site for the structure your student company will build.

Fig. 13-11. Batter boards help mark where a building's foundation will be located.

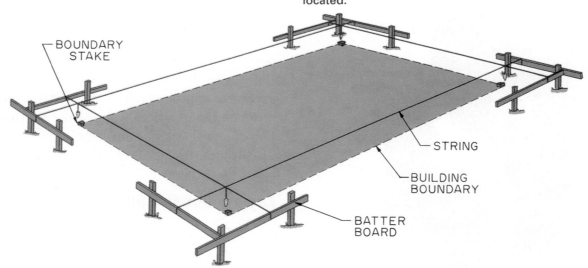

BOUNDARY STAKE

STRING

BUILDING BOUNDARY

BATTER BOARD

Chapter Highlights

● A general contractor manages an entire construction project. A subcontractor specializes in one type of work.

● The project manager uses schedules to keep track of what must be done and when.

● The field office is the center of operations at the construction site. The field engineer and the construction boss work in the field office and oversee the project.

● To prepare the site, several steps must be taken. They include: planning for safety; adding roads, buildings, and utilities; surveying; clearing the land; and laying out the site.

Test Your Knowledge

1. What is the term for a company that manages an entire project?

2. Give an example of what a subcontractor may specialize in.

3. What document sets the maximum cost for materials used on a project?

4. What is the on-site office called?

5. What is the name given to the basic weight of a structure?

6. What is the term for the weight of people and other added items in a building?

7. Name three things that must be done to prepare a site.

8. What workers determine the exact boundaries of the property?

9. What guide does a builder use when clearing the land?

10. What equipment is used to mark a structure's corners on the site?

Correlations

LANGUAGE ARTS

1. Invent a "What Am I?" riddle for any of the nouns in the "Words You Will Need" list on p. 226. The riddle should contain a creative description of the term.

SOCIAL STUDIES

1. Interview an individual who recently worked with a contractor to build a home or business. Ask what the experience was like.

SCIENCE

1. Make a simple sighting device by first gluing a straw to the head of a pin. Fasten a protractor to a cardboard box and poke the pin through the protractor so that the straw will swivel across it. Use the device to set up different angles.

MATH

1. The dimensions of a rectangular construction site are 174' by 360'. How many square feet are included in the site? How many square yards? (Use the formula: Area = Length × Width.)

CHAPTER 14

Construction Processes

Introduction

In Chapter 12, you learned about some of the techniques used for civil construction projects, towers, and pipelines. In this chapter you will learn about the processes involved in residential construction. Although classed as light construction, these processes are similar in many ways to those used in heavy construction.

After studying this chapter, you should be able to

Describe the main parts of a foundation and tell their function.

Describe the main parts of the superstructure and tell their function.

Name the three types of superstructures.

Explain what the four kinds of utility systems are used for and tell how they are installed.

Words you will need

circuit breaker	insulation
climate control systems	load bearing walls
	partitions
drywall	rough work
excavation	sheathing
fixture	siding
footing	studs
foundation	superstructure

The Foundation

The **foundation** supports the rest of a building. Think of it as being everything below the first floor. The remainder of the building is called the **superstructure**. Fig. 14-1. The superstructure will be discussed later.

Excavation

Excavation is the process of digging for the foundation. Most excavations are dug using heavy equipment, such as backhoes and excavators. How much digging must be done depends on the type of construction and the soil conditions.

Excavation also includes rough grading the land. This means shaping the earth to within a few inches of its final slope and elevation. In order to do this, soil may need to be added or taken away. The top layer of soil is usually saved and kept in a corner of the site. This topsoil will be returned to the land when construction is finished. It is better for plants and grass than deeper soil.

The project engineer checks the excavation to be sure the soil is stable and can support the structure. If the soil is not stable, it may have to be packed down. Sometimes additional digging must be done or a different type of foundation used.

Building the Foundation

Two important parts of the foundation are the footings and the foundation walls.

Footings

The first part of the foundation to be laid is the **footing**. A footing is usually made of concrete. It is placed below the foundation wall. Because it is wider than the wall, it helps distribute the weight of the structure over a wider area. Fig. 14-2, pages 238-239, shows a cutaway view of a house. Footings and other important parts of the house are labeled.

The type of footing used depends on the type of construction and the soil conditions. Footings must be below the frost line. The frost line is the depth to which the soil freezes in winter. Fig. 14-2. Freezing and thawing cause soil and concrete to expand and contract. Footings placed above the frost line would be unstable and could harm the rest of the structure.

Fig. 14-1. Everything below the first floor is the foundation. Everything above it is the superstructure.

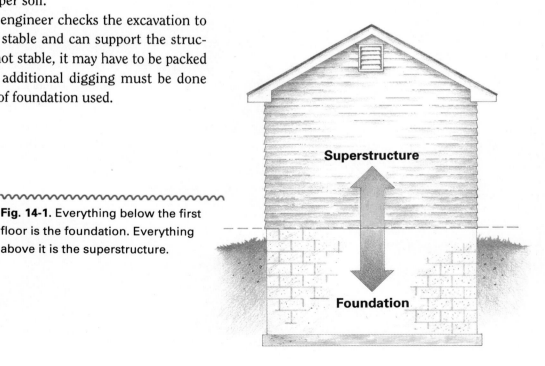

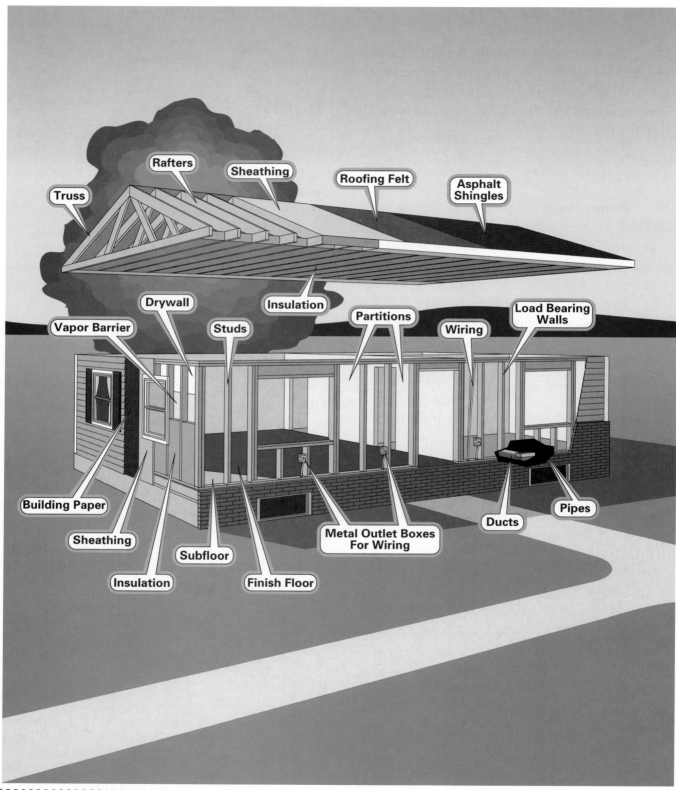

Fig. 14-2. This illustration shows the main parts of the structure of a house. How many do you recognize? Be sure that you learn the purpose of each.

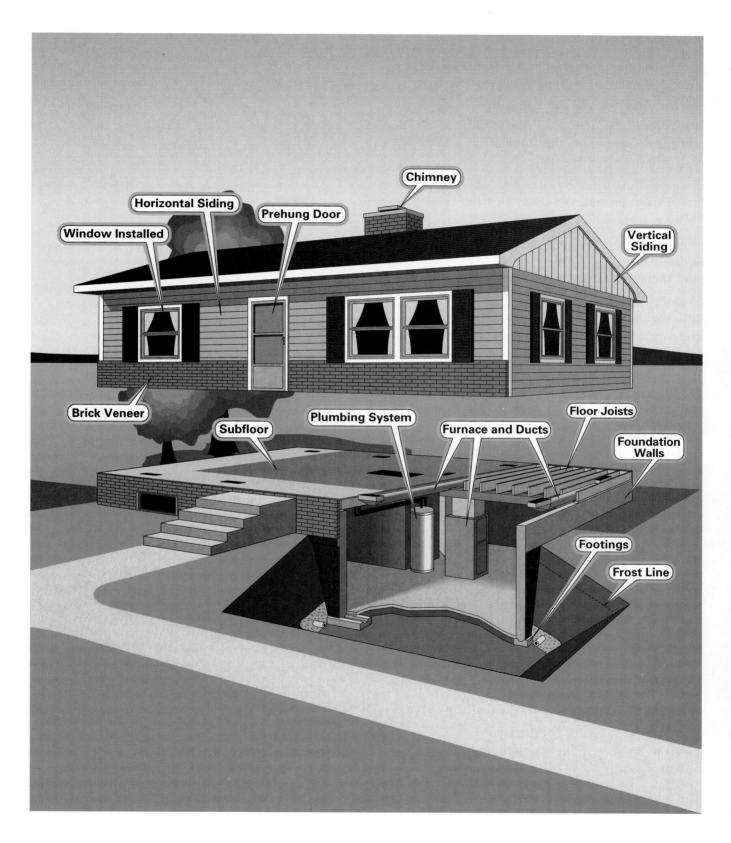

Foundation Walls

The foundation walls are built on top of the footings. Fig. 14-2. They spread the weight of the super-structure to the footings. If the building has a basement, the foundation walls are the basement walls.

Some foundation walls are made from poured concrete. In this process forms (molds) for them are placed on top of the footings. Sometimes steel rods are added to give the walls strength. Then concrete is poured down a chute from a truck into the forms. After a couple of days, the concrete is hard and the forms are removed.

Concrete block can also be used to make foundation walls. This type of wall is built by masons, or bricklayers. The blocks are held together by mortar, a type of concrete paste. Block foundations are less costly than poured foundations.

All foundation walls should be waterproofed. This prevents damage from rain, especially in basements. A thin coating of mortar may be used. Black asphalt, like that used on roofs, is applied next. Plastic sheets or foam insulation may also be required.

Some foundations are made from a single piece of material called a slab. Fig. 14-3. Slab foundations have no walls. The foundation is a concrete slab resting directly on the ground. The superstructure is placed on top of it. There is no crawlspace or basement.

▶▶▶ FOR DISCUSSION ◀◀◀

1. Why do you think concrete is a good material to use for foundations?
2. What do you think would happen to a building if its footings were above the frost line?

Fig. 14-3. This slab foundation rests on the ground. There is no basement.

Extension Activity

■ **Work as part of a group to lay out the foundation for a model of a home. Design and build forms for the footings and foundation walls. Mix and pour the concrete.**

The Superstructure

The superstructure is part of the building above the foundation. It is where we spend most of our time. It has many parts and requires a number of different processes.

Kinds of Superstructures

There are three main kinds of superstructures — frame, bearing-wall, and mass.

Frame Superstructures

A frame for a building is like a skeleton. The many parts of the frame are connected to one another. When the frame is complete, it will support the weight of the building. Later, the frame is covered to form the floors, walls, and roof. Most buildings constructed today have frame superstructures.

IMPACT

Recycling has resulted in new materials and processes for construction. Recycled glass and ground-up automobile tires are being used as ingredients in asphalt for highways. Recycled aluminum and steel are being used to make structural materials, window and door frames, and outdoor lighting fixtures.

Fig. 14-5. This high-rise apartment is framed with concrete.

Steel is used to frame towers, bridges, and large buildings. It is strong and comes in many shapes and sizes. Steel parts, such as beams or girders, are delivered to the site. There, ironworkers weld or bolt them together to form the frame. Fig. 14-4.

Concrete can also be used to frame large buildings, such as hospitals. Fig. 14-5. Forms are built on the site and concrete is poured into them. Sometimes precast concrete frame members are used. These are units formed elsewhere and brought to the site.

Fig. 14-4. This ironworker is welding pieces of a structure.

Fig. 14-6. This building has no frame. The outer walls provide its support.

Fig. 14-7. One of our most famous landmarks, the Washington Monument in Washington, D.C., has a mass superstructure.

Most houses have wood frames. To build a wood frame takes only a few days. Special tools are not required. Standard building materials are used. Installing utilities is easy. These advantages make wood frames popular. The house shown in Fig. 14-2 on pages 238-239 has been framed with wood.

Bearing-Wall Superstructures

Bearing-wall superstructures have no frame. The outside walls, called bearing walls, support the weight of the building. This type of construction is usually used for one- or two-story commercial and industrial buildings. Fig. 14-6. Sometimes the walls are made of concrete blocks or panels. At other times reinforced concrete walls are poured at the site.

Mass Superstructures

Monuments, bridges, and dams may have mass superstructures. Fig. 14-7. A large amount of material, such as concrete, is used to make them. The resulting structures are usually solid, or nearly solid. There is usually little or no usable interior space.

Building the Superstructure

There are many ways to build a superstructure. This section describes some of the basic methods used for floors, walls, ceilings, roof, and sheathing.

Floors

In most houses, the floor is supported by wood framing parts called joists. Fig. 14-2. If there is a second story, the ceiling joists of the first story support the floor of the second.

Several layers of material are usually placed on top of the floor frame. The bottom layer is called the subfloor. Plywood is usually used for this. The top layer is called the finish floor. Fig. 14-2. Finish floors may be made of such things as wood, carpet, or tile.

If a slab foundation has been used, the slab becomes the ground floor. It usually has coverings similar to those used for finish floors.

Walls

Buildings must be protected from the weather. For this purpose, exterior (outside) walls are made. Exterior walls are usually **load-bearing walls**. Fig. 14-2. This means they help support the weight of the structure.

Interior (inside) walls divide the building into different rooms. Some interior walls help bear the load. Others do not. Those that do not bear any load are called **partitions**.

Framing of both interior and exterior walls is usually done with 2" × 4" lumber, called **studs**. Fig. 14-2. Sometimes 2" × 6" lumber is used for exterior walls if thicker insulation will be installed later. Because lumber is becoming more costly, some homes are built with metal framing.

Walls in commercial and industrial buildings are made using many different materials and methods. Concrete and concrete block are often used for bearing walls. Concrete, glass, steel, brick, or wood can be used for exterior walls of framed superstructures. Metal frames are used for interior partitions.

Ceilings and Roof

A roof protects a structure from the weather. The style of roof used depends on the style and function of the building, as well as the load to be carried. Fig. 14-8. For example, in areas that have heavy snows, the roof must have a steep angle so that the snow will slide off. Commercial and industrial buildings often have flat roofs.

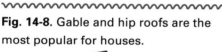

Fig. 14-8. Gable and hip roofs are the most popular for houses.

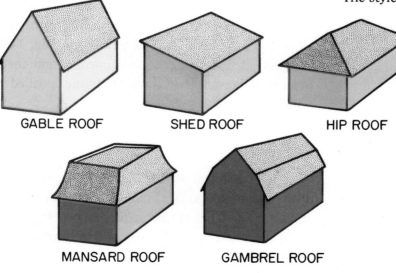

GABLE ROOF   SHED ROOF   HIP ROOF

MANSARD ROOF   GAMBREL ROOF

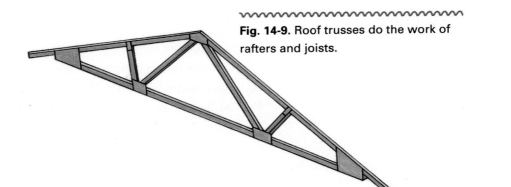

Fig. 14-9. Roof trusses do the work of rafters and joists.

The ceiling and roof in some houses are made using joists and rafters. Rafters are the framing pieces that support the roof itself. They are attached to the ceiling joists. Fig. 14-2. Trusses can also be used. Fig. 14-9. They combine the joists and rafters into one unit and produce a stronger roof. They also cost less and take less time to build.

Sheathing

Sheathing is a layer of material placed between the frame and the outside of a building. Fig. 14-2. Sheets of plywood are often used for sheathing. Wood and other types of siding can be easily nailed to plywood. Plywood also makes the frame more rigid. Sheathing may be applied before or after the roof is framed.

The frame for the roof is also covered with plywood sheathing. The sheathing is then covered with a layer of roofing felt. Fig. 14-2. The felt is a clothlike material combined with asphalt. It helps protect the roof from the weather.

Finishing the Outside

Buildings must have doors and windows to protect them from the weather and to keep heat or cool air inside. They also require an outer covering of some kind and a roof. The edge of the roof and sometimes the corners of the building are enclosed first.

Roofing

Roof coverings include fiberglass, tile, metal sheets, wood shingles, rolls of asphalt, and asphalt shingles. Asphalt shingles are most commonly used on houses.

Prefabricated buildings, which have entire sections made in factories, often have metal roofs. Garages and warehouses are examples. Layers of roll roofing and hot tar covered with gravel are often used for the flat roofs of commercial and industrial buildings.

Doors

Most buildings have more than one entrance for safety reasons. Entry doors are usually made of wood or steel. Steel doors are now very popular. They are well insulated, long lasting, and often attractive.

Most doors are prehung. This means they are already attached with hinges to a frame. Fig. 14-2. Workers set them into the door opening and check to be sure they are level. Then the frame is nailed or screwed into place.

Windows

The architect usually chooses windows as part of the overall design of a building. Windows let light and air inside. They are attractive and help to make a room more pleasant. Popular window designs are shown in Fig. 14-10.

PALLADIAN

BAY WINDOW

OUT-SWINGING CASEMENT

IN-SWINGING CASEMENT

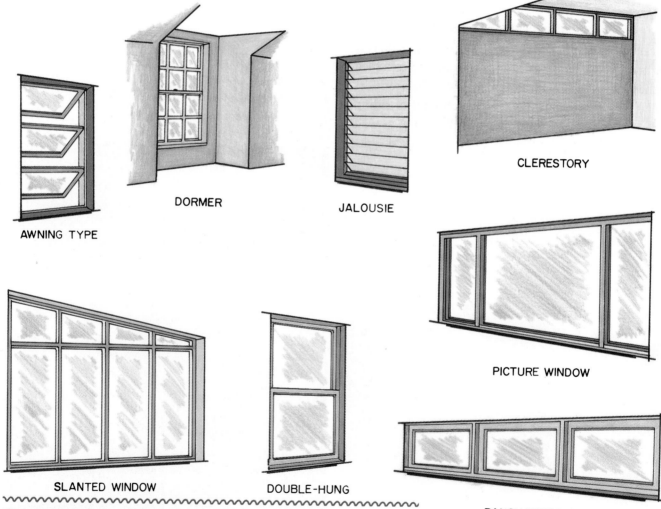

AWNING TYPE

DORMER

JALOUSIE

CLERESTORY

PICTURE WINDOW

SLANTED WINDOW

DOUBLE-HUNG

RANCH WINDOWS

Fig. 14-10. Do you recognize most of these common window designs?

Wood, aluminum, steel, and vinyl (plastic) are used to make window frames. Fig. 14-2. Sometimes wood frames are coated with vinyl to protect the wood from the weather.

The glass in a window is called glazing. Sometimes windows contain as many as three layers of glass separated by a narrow space. The layers are sealed so no air or moisture can get between them. This extra glazing helps insulate the window. Heat or cold cannot go through as easily.

Siding and Masonry

Outside walls may be covered using various materials. Siding usually consists of long, narrow strips of wood, vinyl, or aluminum. Siding is nailed to the studs of a wood frame. Fig. 14-2. The architect of the building usually decides which kind of siding to use. The style and location of the building, as well as the cost of the siding, must all be considered.

Wood strips and boards are used for horizontal or vertical siding. Plywood sheets and wood shingles made from cedar are available in several designs. Vinyl siding can resemble wood and needs no painting. Aluminum siding is also popular and needs little care.

Masonry walls are made of brick, concrete block, and stone. Masonry is fireproof and requires little maintenance. Brick and stone siding are usually installed by a mason. Fig. 14-11. Because they are very heavy, they require more support. The top of the foundation must be wide enough to serve as a base for them.

Fig. 14-11. In A, a worker first installs a plastic layer over the wall. In B, the bricklayer begins to lay the bricks.

A

B

▶▶▶ **FOR DISCUSSION** ◀◀◀

1. **What examples of each kind of super-structure can be found in your community?**
2. **Many new homes have few windows. Why do you think this is so?**
3. **Softwood lumber is widely used for construction. What areas of the country supply the lumber? What would happen if supplies ran low? What if builders began using a substitute for wood?**

Extension Activity

■ **Continue to work with the group that laid out the model home foundation. Frame the floor, walls and roof. Finish a section with sheathing, siding, insulation, and drywall (discussed on following pages).**

Utility Systems

Plumbing, electrical wiring, climate control, and communication systems must all be included in a building. Pipes and wiring go in right after the building is enclosed. This is called **rough work** or roughing-in. Later, after the interior walls are covered, work on the utilities is finished.

Electrical Wiring

Incorrect wiring can create a fire hazard. For this reason, electrical systems should be installed by licensed electricians. In most areas, this is required by law. When wiring is complete, inspectors must approve the job.

Electricity enters a building at the service drop. The service drop may be above or below ground. It is connected to the wiring inside. For roughing-in the wiring, the electrician refers to the building plans. First, the service panel must be installed. The service panel distributes electricity throughout the building. It is also the box that contains the circuit breakers or fuses. Fig. 14-12.

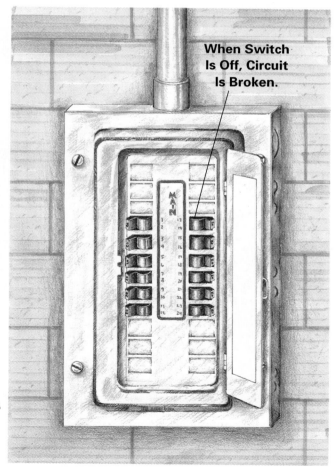

When Switch Is Off, Circuit Is Broken.

Fig. 14-12. The circuit breakers in this service panel protect the building against electrical fires.

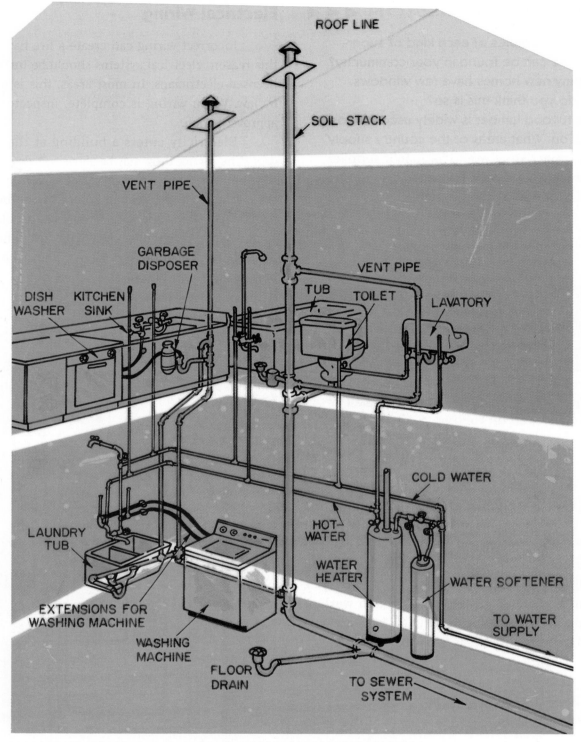

Fig. 14-13. A plumbing system carries fresh water throughout a house and carries away wastes.

A **circuit breaker** prevents wires from overheating and causing a fire. In case of an overload, the circuit breaker opens automatically and stops the current from flowing. A main circuit breaker controls the entire system in a building. Individual circuits may also have circuit breakers. In older buildings, especially homes, fuses perform the same function as circuit breakers.

Next, the electrician mounts metal or plastic boxes for outlets, switches, and fixtures. Fig. 14-2. A **fixture** is a permanent device, such as a ceiling light. Holes are drilled through studs and other parts of the frame. Then wire is run through to each box.

Plumbing

Most buildings need two kinds of plumbing systems. One system supplies fresh water to kitchens, bathrooms, the boiler or water heater, and appliances. The other system carries wastewater and sewage to the septic tank or city sewerage system. Fig. 14-13. Most building codes require that plumbing systems also be inspected when they are finished.

Like electricians, plumbers do their work in stages. In buildings with slab foundations, the rough plumbing is started before the concrete is poured. In other buildings, the rough work is done after the structure is enclosed and before the interior is finished. Fig. 14-2. This is so that pipes can be placed in the framework. The plumber returns to install fixtures, such as sinks, when the interior is finished.

Climate Control

Climate control systems help make a building a healthy and comfortable place to be in. They do this by means of heating, air conditioning, and ventilation (fresh air). Fig. 14-2. Thermostats control the amount of heating or cooling done.

Many forced-air systems combine both heating and cooling. Large pipes called ducts draw the air into an air handler. There it passes over a heating or cooling coil. The conditioned air is then blown through another duct into the different rooms. Fig. 14-14.

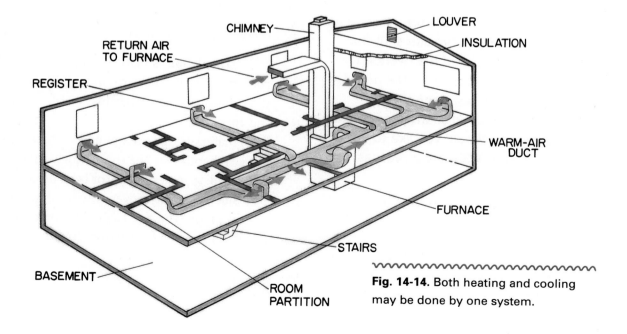

Fig. 14-14. Both heating and cooling may be done by one system.

Hot water systems produce only heat. The water is heated in a boiler fired by electricity, fuel oil, or natural gas. The hot water is then pumped through pipes to radiators in each room. The same boiler may be used to heat water for the tap.

Individual heaters, heat pumps, and solar heating systems can also be used in buildings.

Communication Systems

Communication systems include telephones, cable television, intercoms, and security. Most are installed during construction rather than after. The cables can then be hidden in the walls. Outlets are provided, and items such as telephones or televisions can be plugged in later.

Two common types of security systems are burglar alarms and fire safety systems. These are especially important in commercial buildings. Almost all buildings are now required to have smoke detectors.

▶▶▶ FOR DISCUSSION ◀◀◀

1. Have you ever been present when a fuse was blown or a circuit breaker shut the power off? What caused the overload? What might have happened if the fuse or circuit breaker had not been there?
2. Have you ever used an intercom system? How did it work? Would you like to have one in your home? Give your reasons.

Extension Activity

■ 1. Obtain samples of different types of electrical wire or pipes used in plumbing. Create a display. Label the wire or pipe, telling what material it is made of and what it is used for.
■ 2. Research the history of indoor plumbing. What countries first developed it? When was it introduced into the United States? What was used before it became popular?

The Inside

The last part of a building to be finished is the inside. Insulation is added after the utility systems are roughed-in. Then the ceilings and walls are enclosed.

Adding Insulation

Insulation is material used to keep a building cool in summer and warm in winter. By placing it in the exterior walls, it prevents heat from either coming in or escaping. Fig. 14-2. Insulation is made from materials that are not good conductors of heat.

Fiberglass is the most widely used insulation. Fig. 14-15. It comes in rolls or batts of various lengths. One side is usually covered with aluminum foil or special paper. The foil faces the inside of the house and keeps moisture away from the insulation.

Fig. 14-15. Fiberglass is not a good conductor of heat. This makes it a good insulator.

Walls usually require insulation that is 3½" thick. Ceilings are usually given 6" of insulation. The ideal amount used in a building depends on local weather conditions. For example, a house in Arkansas requires much less insulation than a house in Minnesota.

Insulating materials are labeled according to their R-value. The higher the R-value, the better the insulating qualities. For example, 3½" fiberglass insulation has an R-value of 11. Six-inch fiberglass has an R-value of 19.

Covering Walls and Ceilings

In years past, walls and ceilings were covered with plaster, a heavy paste which hardened. Today, the most common material used is **drywall**. Drywall comes in large sheets and consists of a plaster-like filler covered with paper. Fig. 14-16. After it is installed, drywall can be painted or covered with wallpaper.

Ceilings are done first. The large sheets of drywall are nailed or screwed to ceiling joists. Then the walls are done. The drywall is fastened to the studs. Fig. 14-2.

Next, any nail holes or seams are filled with joint compound. Joint compound is similar to plaster. Tape is pressed into the seams and more compound is applied. After this is dry, two more coats of joint compound are added. With light sanding, this leaves a smooth, even surface.

After the drywall is installed, strips of material called molding are added for decoration. Molding is placed around the doors, windows, and floor. Any cabinets are installed. The inside of the building can then be decorated with paint or wallpaper.

Fig. **14-16.** Drywall is used to cover interior walls.

►►► FOR DISCUSSION ◄◄◄

1. Look at a map. Find these states and discuss the amounts of insulation that buildings there might need: Montana, North Carolina, Maine, Indiana, South Dakota, Florida.

2. Have you ever helped paint or wallpaper a room? Describe your experience.

Company Activity

■ Suppose the structure your student company is building has to be insulated. Research the insulation requirements in your area. Next, visit a store that sells insulation and compare brands and prices. Then recommend to the company the best kind of insulation to use.

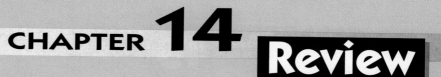

Chapter Highlights

● The foundation supports the rest of a structure. Excavation is the process of digging the foundation. Footings and foundation walls are the most important parts of the foundation.

● The superstructure includes everything above the foundation. The three main kinds of superstructures are frame, bearing-wall, and mass. Basic parts of the superstructure include floors, walls, ceilings, roof, and sheathing.

● Roofing, doors, windows, and siding are all necessary for finishing the outside of a building.

● Utility systems needed in a building include plumbing, electrical wiring, climate control, and communications.

● The inside of a building is finished by adding insulation and covering the walls and ceilings.

Test Your Knowledge

1. What is the purpose of a foundation?

2. Name the two main parts of most foundations.

3. List the three main kinds of superstructures.

4. What is the most popular material for framing a house?

5. When interior walls do not bear any load, what are they called?

6. What is the layer of material placed between the frame and the outside of the building called?

7. Name three kinds of roof coverings.

8. What are the four utility systems that are included in buildings?

9. What material is the most widely used for insulation?

10. Name the most common material used to cover walls and ceilings.

Correlations

LANGUAGE ARTS

1. A metaphor compares two unlike things by saying that one *is* the other. For example, the **frame** of a building *is* its **skeleton**. Write several metaphors comparing other parts of a structure to the different parts of the human body.

SOCIAL STUDIES

1. Watch the processes involved in a residential construction. Ask older workers how materials, quality of construction, time management, and safety precautions have changed.

SCIENCE

1. Draw a circuit diagram that has three light bulbs connected in *series* with a battery. Draw another diagram with three light bulbs connected in *parallel* with a battery. When one bulb is unscrewed in each of the circuits, what will happen to the other two?

MATH

1. One bundle of fiberglass shingles will cover 35 sq. ft. Shingles are $5.95 a bundle. If your roof is 50' by 70', how much will shingles cost?

CHAPTER 15

Controlling Construction Quality

Introduction....................................

In 1981, a suspended walkway inside a large hotel in Kansas City, Missouri, collapsed during a dance. One hundred fourteen people were killed and 185 were injured. Why was this building unsafe? Who was responsible? (To learn more about it, see the Impact feature on page 256.)

As with manufactured products, construction projects must also meet quality standards. The goal of quality must be established before construction begins. Rules and regulations must be followed. The architects, contractors, workers, public officials, and other people involved must all do their jobs to the best of their ability.

After studying this chapter, you should be able to......................

List the goals for a construction project.

Explain the purpose of zoning limits, building permits, and building codes.

Give examples of how various workers contribute to construction quality.

Explain how materials, methods, and work performance are inspected and describe final inspections.

Words you will need ....

building codes
building permit
certificate of occupancy
punch list

The Project Goal

Control of construction quality begins when the project is first proposed. It continues during design and actual building. The goal of quality control is to produce a structure that

- is of acceptable quality
- is finished on schedule
- conforms to all laws
- is built for an amount equal to or below the estimated cost.

Many different activities take place in order to achieve the goal. These activities involve giving instructions, checking the work, and adjusting the work as needed. For example, supervisors and owners tell workers what to do. The different contractors check one another's work. Fig. 15-1. If problems are found, changes are made.

Fig. 15-1. This electrician cannot complete a wiring job properly because the plumber installed pipes in the wrong place. The problem will have to be reported to the supervisor.

▶▶▶ FOR DISCUSSION ◀◀◀

1. **Think about the universal systems model. Where do quality checks fit into that model?**

Company Activity

■ **What are the goals of your student company? Individual class members should take a minute to write them down. Be specific. For instance, instead of saying, "The work should be done on time," give the exact date. Now, as a group, make a list of the goals on the blackboard. Was everyone's list of goals the same? If not, what problems can you foresee when you work together on your project? Why do you think it is important for people working together to have a clear understanding of what the goals are?**

Rules and Regulations

Buildings should be safe for people to use. They should also provide adequate shelter. Windows and doors should open and close properly. Roofs should keep out rain and snow. The plumbing and other utilities should work as expected.

To make sure buildings are built correctly, governments set up a series of rules and regulations. These include zoning rules, building permits, and building codes. They help to ensure quality in construction.

IMPACT

Poor construction quality can kill. Investigators who studied the collapse of a walkway in a hotel in Kansas City found building code violations. Improper materials had been used. Welds fractured and could not support normal loads. The hotel had been in service for only one year.

Zoning

As you know, zoning rules limit construction. They establish where certain types of structures can be built. For example, zoning laws prevent factories from being built in residential neighborhoods.

Zoning rules also limit where buildings can be located on a site. A building must be set back a certain distance from the street and from the property lines. Fig. 15-2. There may also be a limit on how much of the lot the building can cover. Sometimes buildings may have to be a certain minimum size.

Building Permits

Most communities require builders to get a **building permit** before construction begins. The permit is proof that construction has been approved by the city or county.

First, the building department reviews the plans. Sometimes the architect or engineer is asked to make changes in the plans to satisfy local regulations. Then the permit is issued. The permit must usually be displayed at the site until the structure is finished. Fig. 15-3.

Building permits help control the quality of construction. They also help make sure the building conforms to local laws.

Fig. 15-3. Building permits must be posted at the construction site before work begins.

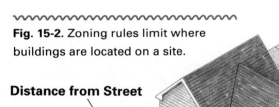

Fig. 15-2. Zoning rules limit where buildings are located on a site.

Distance from Street

Minimum Distance from Property Lines

Street

Building Codes

Building codes are set up by state and local governments to protect the owners and users of buildings. As you learned in Chapter 4, the codes specify the materials and methods that can be used for construction. The standards required by the codes are based on the main use of a building. For example, only two exits may be required for a single-family home. A large movie theatre may be required to have four.

The architect or engineer who designs a building is usually aware of all the codes that apply. He or she accepts responsibility for making sure the building conforms to those codes.

Most state codes are based on national standards, such as the National Electrical Code. Local codes usually follow these standards as well. However, sometimes local codes are more strict.

During construction, a building inspector visits the site to make sure the codes are being followed. If the inspector sees a problem, a written notice is sent to the owner. Later the inspector returns to make sure the problem has been corrected.

Company Activity

■ Obtain a copy of the building code that applies to the structure your company is building. Make a list of specifications for your structure that conform to the code.

The Quality Control Team

Many people are involved in controlling construction quality. They include public officials, architects, engineers, owners, contractors, suppliers, and lenders. Following are some examples of the jobs they do.

Owners are perhaps the most concerned with quality. Usually owners have a large financial stake in the project. If the building has defects, they will have to deal with the problems. They want to make sure the project is what was promised. If they are unhappy with the work being done, they should discuss their concerns with the contractors.

▶▶▶ FOR DISCUSSION ◀◀◀

1. Do you know of cases in which people in a neighborhood objected to a new building project? If so, what happened?
2. In most communities, a building permit must be displayed at the construction site. Why do you think displaying the permit is necessary?

IMPACT

Structures built to solve one problem may cause others. Egypt is a desert country. During the dry months, droughts occur. During the rainy seasons, there are floods. To prevent these extremes, the Aswan High Dam was built. As a result, flooding has been controlled. Water is available during dry periods. However, the dam has damaged the environment. The soil used to be enriched by the floods. Now farmers must use fertilizers. Erosion is also a problem. Some diseases have increased because flood waters no longer wash away dangerous organisms.

Fig. 15-4. If supplies reach the site on time, work stays on schedule.

Architects and engineers have designed the project. They have attempted to follow all the building codes that apply. During construction, they visit the site to be sure their plans are being followed.

Contractors make sure that the people who work for them do their jobs correctly. A general contractor checks the work of subcontractors who have been hired for such jobs as plumbing and electrical wiring. In addition, subcontractors will also have their work checked by their supervisors. At many jobs, a superintendent oversees the entire project.

Suppliers also contribute to quality. They make sure the correct materials are shipped at the right time. Fig. 15-4. If mistakes occur, projects are delayed and contractors can lose money. Suppliers may also lose money if they have to pay for additional deliveries.

Banks that lend money for a project do not usually give the money in a lump sum. Instead, they give it in stages as the project moves along. They keep an eye on construction and set a schedule for payments. For example, they might give one-third of the loan when excavation begins. Another third is paid when the structure is framed. The rest is given when the structure is enclosed.

OSHA inspectors visit sites to see that federal laws are followed. They also make sure the site is safe for workers. Insurance company representatives inspect sites insured by their companies. They look for dangerous conditions that might cause accidents.

Almost every town has at least one building inspector. Small towns may hire a part-time inspector. Large cities may hire dozens of workers full-time. The inspector makes sure that at least minimum standards are being met on the project.

▶▶▶ FOR DISCUSSION ◀◀◀

1. **Contractors must make sure that workers do their jobs properly. What could contractors do to encourage employees to take pride in their work?**

_____

Company Activity

■ **When the structure your company is building is finished, ask a local building inspector to check it over. What did the inspector look for? How well did your company do?**

Inspections for Quality

Inspections begin almost as soon as construction starts. They continue until the structure is ready to be used. Materials, methods, and work performance are all checked.

Materials

The architect and owner usually choose the building materials. Building codes, costs, and personal preferences are all considered. Then the architect writes the specifications. The specifications include many details about the materials that should be used. When purchased, these materials must be as good or better than those called for in the plans and specifications.

Some materials are easy to inspect. For example, lumber is graded and stamped at the mill. Fig. 15-5. When the shipment of lumber is delivered, someone can easily check the stamp against the specifications.

Fig. 15-5 Look at the stamp on this lumber. "S-P-F" means it is spruce, pine, or fir. "Dry" means it has been dried in a kiln. "No. 1" indicates that it is an all-purpose grade of lumber.

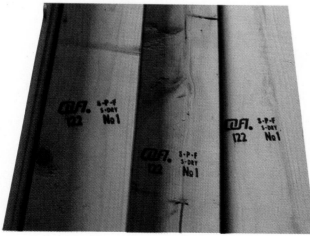

Other materials can be checked through testing. For example, to find out if the soil at a site is stable enough, samples can be taken. The samples are then tested at a laboratory. Concrete delivered to a site each day may also be tested. The tests determine whether or not the proportions of cement, gravel, sand, and water match what the specifications call for. Tests are necessary because a poor mixture will not be strong. Important concrete parts of the structure could crack or collapse.

Methods

Methods used during construction are also inspected. Footings are checked to make sure they are below the frost line. Foundation walls are checked to be sure they are thick enough.

During roughing-in, electrical work is inspected to make sure the right type and number of circuits have been installed. Plumbing is inspected to be sure the angle of drainpipes is correct. If it is not, waste will not flow properly to the septic or sewerage system.

Work Performance

Quality materials and methods cannot guarantee a good job. The work itself must be of good quality. Sloppy work may result in such problems as leaky roofs, cracks in walls, and doors that do not close properly. To assure quality work, owners must choose a contractor carefully. The contractor should have some experience and a good reputation. It is also a good idea to talk to people for whom the contractor has worked.

The job superintendent must set high standards and see that they are met. Supervisors who oversee the workers must be able to recognize the difference between good and poor quality work. Workers should be encouraged to take pride in their work. Owners and contractors can help by providing a good working environment.

Final Inspections

Most structures undergo final inspections by several people. One is done by the building inspector. Fig. 15-6. When any requested adjustments have been made and all main systems work properly, the inspector issues a **certificate of occupancy**. This certificate says that the building is ready and safe for people to use.

Another final inspection is made by representatives of the owner, architect or engineer, and by the contractor. Every detail is checked. Every system is tested to make sure it works. Any corrections that need to be made are listed. This list is called a **punch list**.

The punch list is given to the contractor. A reasonable deadline for making corrections is set. When the corrected work is complete, the inspection team returns. The building is finished when everything on the punch list has been corrected. If problems still remain, further repairs and inspections will be required.

Extension Activity

■ **Make two forms into which you will pour concrete. The forms should measure 1½" x 1½" x 14".** Follow your teacher's instructions for mixing two batches of concrete. One batch should be mixed using the proper proportions of materials. The other should contain too much water. Let the concrete dry. After seven days, compare the strength of the two samples. Use a commercial testing machine or one which you design. Follow all safety precautions set up by your teacher.

▶▶▶ FOR DISCUSSION ◀◀◀

1. Suppose you are the owner of a new house that is being built. You go to the building site to make a final inspection. What are the most important things you will look for?

2. What are the advantages and disadvantages of using materials of higher quality than those required by building codes?

Fig. 15-6 A building inspector checks the structure to be sure it meets all building code requirements.

Chapter Highlights

● The goal of quality is essential to every construction project.

● Rules and regulations for construction include zoning rules, building permits, and building codes.

● The quality control team includes public officials, architects, engineers, owners, contractors, suppliers, and lenders.

● Inspections for quality begin with construction and continue until the structure is ready for use. Materials, methods, and work performance are all checked.

● Final inspections are performed by the building inspector, the contractor, and representatives of the owner, architect, or engineer.

Test Your Knowledge

1. Which rules limit where buildings can be built?

2. What document is used to show that construction has been approved by the city or county?

3. Whom do building codes protect?

4. List four people involved in controlling construction quality.

5. What federal agency checks a site for worker safety?

6. When do inspections for quality begin?

7. Who usually chooses the building materials?

8. What is the name of the document that says a building is ready and safe for people to use?

9. Name two people who make final inspections.

10. What is the list of corrections made during final inspection called?

Correlations

LANGUAGE ARTS

1. Write an expository paragraph explaining your career goal and the type of training or education you will need to reach that goal.

SOCIAL STUDIES

1. Some countries, such as England, take more care than we do to preserve examples of fine architecture. Find a building in your town that you think should be saved. List your reasons.

SCIENCE

1. Write a report describing how a circuit breaker and a fuse work in electrical circuits. How are they similar? How are they different? What are the advantages and disadvantages of each?

MATH

1. A bank will loan a builder $450,000. It will give one-third of the loan when excavation begins. How much will the builder receive?

CHAPTER 16

After Construction

Introduction.....................................

Many activities take place after construction. Finishing touches must be applied to the structure. The site must be landscaped. As the structure is used, it must be maintained and repaired. Later, the owner may decide to make changes in it. In this chapter, you will learn about some of the ways structures are altered to look and function better.

After studying this chapter, you should be able to......................

Describe ways in which a building's exterior and interior are finished and decorated.

Explain how landscaping is done and tell the difference between constructed features and plantings.

Tell the difference between maintenance and repairs and explain why they are necessary.

Describe remodeling and restoration and give examples of each.

Words you will need ....

constructed features
flat paints
gloss paints
grading
landscaping
maintenance
plantings

preventive
    maintenance
primer
remodeling
repairs
restoration
semigloss paints

Finishing and Decorating

Finishing and decorating begin when construction is complete. Work must be done both outside and inside the structure.

Outside

After a general cleanup, work is done on outside walls, windows, doors, and trim. Walls of brick, stone, vinyl, or aluminum siding need no finishing. They also require little maintenance later. Wood, however, must be protected from the weather. Fig. 16-1. Some wood siding is finished with a clear preservative. Other types of wood may be primed and painted. The **primer** is a coating that seals the wood. The paint adds color.

If windows and doors are of bare wood they must be protected from the weather. People who prefer the look of natural wood may apply a clear finish. Clear finishes, such as weatherproof varnish, let the wood grain show through. The wood can also be stained first, which adds some color. Other people prefer the look of paint. In this case, the wood is given both a primer and a finish coat of paint.

Wood trim is also primed and painted. Shutters are an example. The paint protects the wood from the weather. It may also be used to add a contrasting color for decoration.

Inside

Before finishing can be done on the inside, waste, dust, and dirt must be removed. The floor is swept, and the walls and ceiling are cleaned. Any built-in cabinets are installed and decorating is done.

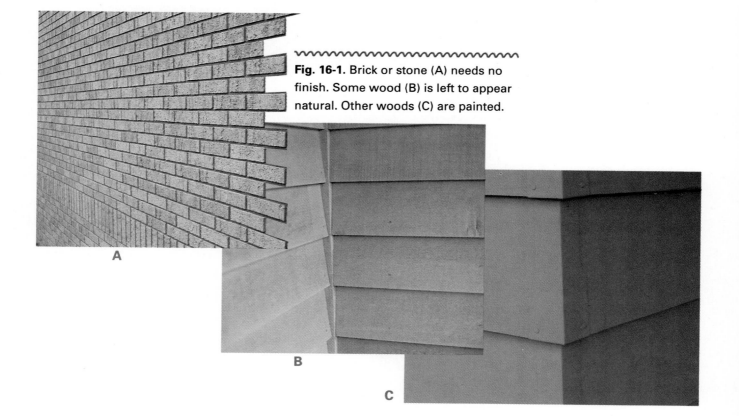

Fig. 16-1. Brick or stone (A) needs no finish. Some wood (B) is left to appear natural. Other woods (C) are painted.

A

B

C

Cabinets

Cabinets are usually found in the kitchens and bathrooms of most homes. Fig. 16-2. They are used for storage. Cabinets may be made of metal, wood, or wood products, such as particleboard. Some cabinets can be bought prefinished. Unfinished cabinets can be painted, stained, and/or varnished.

Ready-made cabinets are available in many sizes and styles. Fig. 16-3. Customers generally select them from a catalog or visit a showroom. Some dealers use computers to show customers the different ways cabinets can be arranged. Cabinets can also be custom-made by cabinetmakers.

Fig. 16-2. Built-in cabinets are made to fit in a certain spot. They may be made of metal, wood, or wood products.

Fig. 16-3. Ready-made cabinets come in different styles. What are the differences among those shown here?

Fig. 16-4. Which of these two paints can be cleaned up with soap and water? Which requires turpentine?

Paint

Almost all buildings have painted areas. Paint is the most popular finishing material for interiors. Compared to other finishes, it is inexpensive. It is easy to apply and helps protect a material's surface. Also, paint quickly gives the structure a fresh, new look. Painted surfaces are also fairly easy to clean.

Paint may be used to make a surface dull, shiny, or somewhere in-between. **Flat paints** leave a dull surface. **Gloss paints** leave a very shiny surface. **Semigloss paints** are softer with only a little shine. A flat finish is usually used on walls and ceilings. The flat finish helps reduce glare from lights. Semigloss finishes are often used for trim, doors, windows, and kitchen and bathroom walls. They wear well and can be washed often. Sometimes gloss paint is also used for trim.

Most paints have either a water or an oil base. Water-base paints, such as latex, dry quickly and can easily be washed off brushes with soap and water. Oil-base paints take longer to dry. Brushes must be cleaned with turpentine or paint thinner. Fig. 16-4.

Water-base paint is usually used for walls and ceilings. Oil-base finish may be used for trim work. First, the primer is applied. This coat helps seal the unfinished surface. The surface is then prepared for the finish coat. If the finish coat is to be a water-base paint, then the primer must be water-base as well. The same applies to oil-base paints.

Paint is usually applied with brushes and rollers. Rollers cover large areas quickly. Brushes work best in corners and on edges and trim. A fairly new paint tool, the paint pad, has become popular. The pad is flat and has a handle. The pad's flat surface is loaded with paint and swept across the wall surface. Like rollers, pads are fast and efficient. In commercial and industrial buildings, paint sprayers may be used.

Wallpaper and Paneling

Many people prefer walls finished with wallpaper or paneling. Wallpaper comes in thousands of designs and is popular in homes and commercial buildings. Today, the term wall*paper* is no longer entirely accurate. Although some wallpapers are still made of paper, many are vinyl or vinyl-coated. Vinyl wallpaper is washable. For that reason it is preferred for kitchens and bathrooms.

Paneling can be made of real wood or synthetic materials. It comes in many styles. Paneling can be fastened directly to wood or metal studs. Using nails or a type of glue, it can also be applied over finished walls.

Floor Coverings

Floors can be finished with hardwood, vinyl flooring, tile, or carpeting. Hardwoods such as maple and oak are usually used for floors. The wood is cut into strips. These strips are then nailed or glued to the subfloor. Unfinished floors are usually sanded and then protected with several coats of clear finish. Prefinished wood flooring is also available.

Vinyl flooring is sold in large sheets six or twelve feet wide. Most rooms can be covered with a single piece that is cemented into place. Vinyl is popular for kitchens, bathrooms, and heavy traffic areas.

Tile flooring is also used for heavy traffic areas. Fig. 16-5. Ceramic tile is very durable. It is used in kitchens and bathrooms. In commercial buildings it may also cover the walls as well as the floors. Tile may be made of vinyl, wood, and other materials as well.

Carpeting can be laid over the subfloor or a finish floor. Workers who specialize in laying carpets usually do the job. Carpeting is popular in homes, offices, and public areas of industrial and commercial buildings.

Look at Fig. 16-6. Try your skill at estimating the materials needed to finish the room shown.

▶▶▶ FOR DISCUSSION ◀◀◀

1. What materials have been used on the outside walls of your school? What do you think the advantages and disadvantages of these materials are?
2. Many designers believe that colors affect people's moods. What colors would you paint a sunny room? A dark, shadowed room? A kitchen? A walk-in closet? Give your reasons.
3. Have you or someone in your family ever tried putting up wallpaper? Describe what happened.

_____

Fig. 16-5. Tiles are laid one at a time. They provide a lasting surface.

How Much Will You Need?

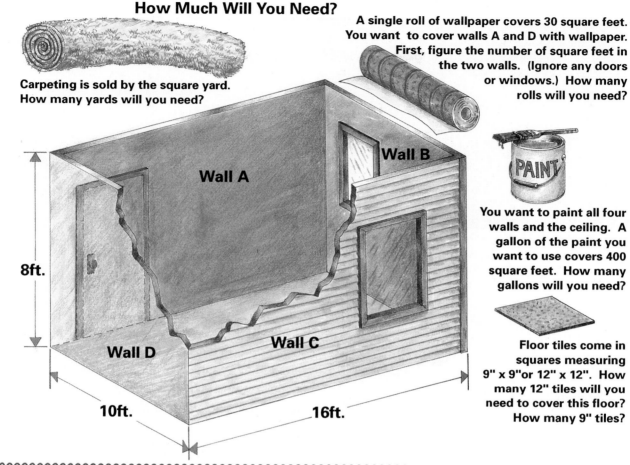

Carpeting is sold by the square yard. How many yards will you need?

A single roll of wallpaper covers 30 square feet. You want to cover walls A and D with wallpaper. First, figure the number of square feet in the two walls. (Ignore any doors or windows.) How many rolls will you need?

Wall A

Wall B

Wall C

Wall D

8ft.

10ft.

16ft.

You want to paint all four walls and the ceiling. A gallon of the paint you want to use covers 400 square feet. How many gallons will you need?

Floor tiles come in squares measuring 9" x 9" or 12" x 12". How many 12" tiles will you need to cover this floor? How many 9" tiles?

Fig. 16-6. Using the information given, estimate the materials needed to finish this room.

Extension Activity

■ **Suppose you had permission to redecorate your room at home. Make some simple sketches of the room as you would like it to be. Indicate colors and fabrics. Write a brief paragraph telling how you would go about selecting the materials you needed.**

Landscaping

Landscaping is the development of the land and its features. This provides a better environment for people to live and work in. Landscaping not only improves the appearance, but it adds value to a building. If the project is small, a landscape plan is drawn by the owner, the building architect, or a landscaping firm. Large projects are designed by professionals called landscape architects.

The plan shows the shape of the earth, constructed features, and plantings. **Constructed features** are those that must be built, such as decks,

Fig. 16-7. By following this plan, workers know where to place constructed features and plantings.

driveways, and fences. They add to the function and appearance of the building but are not really part of it. **Plantings** include any trees, bushes, grass, and other plants that surround the building. Fig. 16-7.

Before landscaping can begin, trash left from construction must be cleared. Trenches dug for water pipes, gas lines, sewer drainpipes, and electric or telephone cables must be covered. When the soil is put back into these trenches, the process is called backfilling.

Next, the soil must be graded. **Grading** gives the earth the shape called for in the landscape plan. Any topsoil that was removed during excavation is usually returned at this time. As a rule, the soil is made to slope away from the structure. This prevents rainwater from collecting around the foundation. The slope is gradual so that the soil will not wash away.

Constructed Features

Constructed features are often made of concrete, lumber, asphalt, or brick. The material chosen depends on cost and the look desired. Some of the same people who helped make the structure may also work on constructed features.

Walks and driveways are the most common constructed features. Fig. 16-8. Most of these are

Fig. 16-8. Most buildings require a sidewalk leading up to them. The concrete for this walk has been poured and the surface is being smoothed.

made of poured concrete. Patios and decks are also popular. Most patios are made of concrete or brick. Decks are constructed from weather-resistant wood. Cedar, redwood, and pressure-treated lumber are often used. Pressure-treated lumber is not damaged by moisture or insects and can last 30 to 40 years. Decks often include other features such as benches, lights, and planters (plants in containers).

Retaining walls may be added to hold back soil so it doesn't wash away. They can be made of such materials as concrete and pressure-treated lumber. Some fences are used for privacy and security. Others are only for decoration. Fences can be made of wood, brick, stone, or metal.

Other constructed features include fountains, ponds, swimming pools, benches, and outdoor lighting.

Plantings

After the topsoil has been replaced, plantings can be put in. Trees are usually planted first. Heavy equipment may have to be used to move them or dig the large holes they require. If trees already exist on a site, efforts are usually made to keep them.

Bushes and shrubs are added next. They come in many sizes, shapes, and colors. They also bloom at different times of the year. Landscape designs try to take advantage of these differences. For example, low shrubs would be planted in front of taller ones. A variety might be chosen to make sure at least one is in bloom at any given time of the year. Fig. 16-9. Holes for bushes and shrubs are usually dug by hand.

Fig. 16-9. In the earlier picture, the topsoil has been graded. Later, sod and plantings were added.

Flowers can be planted at the same time as shrubs. They are the smallest and most colorful plantings. Some flowers bloom for many seasons. Others must be replaced every year. This helps to provide the landscape with a new look at a low cost.

Grass is usually planted last. Stones, roots, and other debris are raked away. Then grass seed or sod is planted. Sod is a layer of soil with grass already growing on it. It creates an "instant" lawn.

▶▶▶ FOR DISCUSSION ◀◀◀

1. Scientists are developing new varieties of grass that will never need mowing. What impacts will this type of grass have on the economy? On society?

Company **Activity**

■ **Develop a landscape plan for the structure your company is building. Using symbols you create for trees, shrubs, flowers, and grass, make a drawing of your plan.**

Maintenance

Maintenance refers to taking care of a structure and the surrounding property. Maintenance keeps things running smoothly. It continues throughout the life of the structure.

Cleaning, vacuuming, and removing waste are regular maintenance activities in most homes and other buildings. Replacing lightbulbs and washing floors and windows are others. They are performed on a regular basis.

Landscaping also needs regular maintenance. Many people water their lawns during hot summer days. Some use fertilizers. Bushes and trees must be trimmed occasionally. Weeds must be removed from flower gardens, and some flowers must be replaced.

Preventive maintenance is done regularly to prevent problems from occurring. For example, filters on forced-air furnaces must be changed on a regular basis. Clean filters keep the system working efficiently for a longer time. Repainting siding is also a form of preventive maintenance. With a new coat of paint, the siding is protected from moisture and lasts longer. Fig. 16-10.

Structures such as bridges may require full-time maintenance workers. For example, long suspension bridges are often being continually repainted. The painters begin at one end of the bridge. They scrape and give rusted areas a coat of primer. Then they apply a finish coat. When they reach the opposite end of the bridge, they begin the process again. This constant repainting greatly reduces damage from the weather. Maintenance costs are also reduced in the long run.

Most homeowners take care of many of their own maintenance jobs. This saves money. A maintenance staff may be hired to take care of commercial, industrial, and large residential buildings. For special jobs, such as exterior painting, outside contractors may be called in.

Repairs

Repairs restore a structure to its original condition. Almost all structures need repairs at some point. Minor repairs can be taken care of quickly and at reasonable cost. Major repairs, such as giving a highway a new surface, can cost millions of dollars and take months to complete.

Fig. 16-10. This house now needs many repairs. What kinds of maintenance could have been done to prevent problems you see here?

Replacing the weatherstripping around a door is an example of a minor repair. The weatherstripping helps seal the door against the weather. It saves money in heating and cooling expenses. Weatherstripping is inexpensive and takes little time to install. Fig. 16-11. Replacing washers in leaky faucets is another minor repair that results in savings. More serious problems, such as a leaky roof, must be corrected to prevent costly damage to the interior.

Sometimes it is better to replace an item rather than repair it. Parts may not be available. If the structure is old, other problems may soon occur. Replacement prevents these problems.

For commercial, industrial, and large residential buildings the maintenance staff often takes care of routine repairs, such as clogged drains. Outside contractors may be hired for specialized jobs.

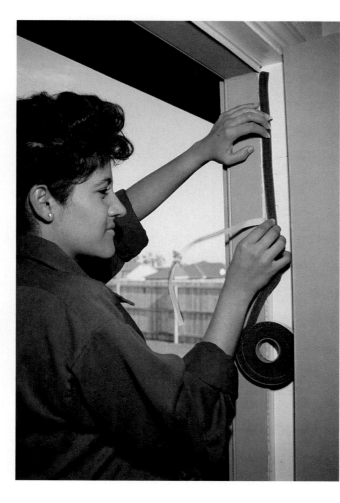

Fig. 16-11. Weatherstripping helps keep out heat or cold.

►►► FOR DISCUSSION ◄◄◄

1. What kinds of problems do you think a leaky roof could cause? A leaky basement? A broken storm window?
2. Some manufacturers offer service contracts on large appliances such as water heaters. Under the contract, the manufacturer takes care of any repairs at no additional cost to the owner. Do you think such contracts are a good buy? Give your reasons.

Extension
Activity

■ **Design a test to compare the toughness of two different kinds of flooring. Conduct the test in a high traffic area of your technology laboratory. Record your observations on a regular basis. Which type requires less maintenance?**

Alterations

Alterations are changes made to a structure's form. Alterations can change its size or make it more modern. Some alterations are done to change the way in which a building is used. For example, a hair stylist might want to make part of his or her home into a beauty shop. Not all alterations are done to buildings. Adding more lanes to a highway is an example of altering a different kind of structure.

The two types of alterations are restoration and remodeling.

IMPACT

In cities, urban renewal projects can save historic buildings. They can also increase the supply of affordable housing.

Fig. **16-12.** The town of Williamsburg, Virginia, was restored to its original state. Millions of tourists visit it each year. The people who work in Williamsburg dress as people did in the 1700s.

Restoration

Restoration returns a building to the way it looked originally. Another term for restoration is renovation, which means "to renew." Historic buildings that are good examples of architectural design are often restored. Williamsburg, Virginia, is an example of an entire town that was returned to the way it looked in the 1700s. Fig. 16-12.

There are historic districts in some communities. Changes to the outside of historic buildings can only be made with permission. However, work on the inside and maintenance on the outside are usually permitted.

Some unusual structures are also restored. In 1885, the 151-foot-high Statue of Liberty was presented to the United States from France. One hundred years later, it needed many improvements.

The statue's copper skin was cleaned and damaged areas were repaired. The inside of the structure was made more comfortable for visitors. The spiral stairs were given new treads to make climbing easier and safer. New lighting and a new heating and cooling system were installed. The most important improvements made the statue a better and stronger structure. Fig. 16-13. The cost of restoring the statue was $62 million. All of this money came from donations.

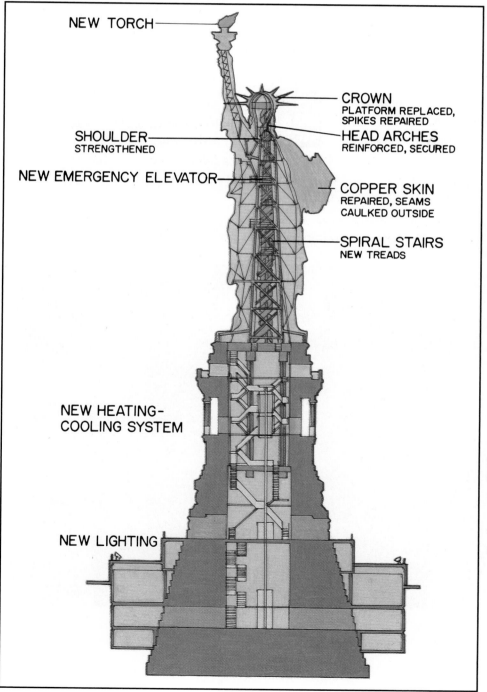

NEW TORCH

CROWN
PLATFORM REPLACED,
SPIKES REPAIRED

SHOULDER
STRENGTHENED

HEAD ARCHES
REINFORCED, SECURED

NEW EMERGENCY ELEVATOR

COPPER SKIN
REPAIRED, SEAMS
CAULKED OUTSIDE

SPIRAL STAIRS
NEW TREADS

NEW HEATING-
COOLING SYSTEM

NEW LIGHTING

Fig. 16-13. The Statue of Liberty was restored for her 100th birthday. This cutaway view shows how structural reinforcements were added.

Remodeling

Remodeling is done to make a building more modern and add to its usefulness. For example, air conditioning or an enclosed porch might be added.

Remodeling is often done by new owners. They buy an older building and make it more suitable for their needs. A family with children might add another bedroom. An attached garage might be made into a family room. A basement might become a game room.

Remodeling adds value to a home. A new kitchen or bathroom can make it more modern. Fig. 16-14. Replacement windows and doors or a new heating system can make a house more comfortable and save money later on. Such alterations as air conditioning and skylights are not absolutely necessary. However, they do make the home a more enjoyable place to live.

▶▶▶ **FOR DISCUSSION** ◀◀◀

1. Suppose your family chose to remodel a home rather than move. What reasons might they have had for such a choice?
2. Restoring old structures is usually more expensive than replacing them. What advantages are there to restoring historic buildings?

Extension Activity

■ Photograph a building or other structure that is being altered. Display the pictures in an organized way. Write a caption for each picture.

Fig. 16-14. How many improvements can you find in the remodeled version of this room?

Chapter Highlights

● When construction is complete, finishing takes place both inside and outside the structure. Outside finishing includes work on outside walls, windows, doors, and trim. Inside finishing includes putting in cabinets, painting, covering walls, and covering floors.

● Landscaping develops the land and its features.

● Constructed features and plantings are used in landscaping.

● Maintenance refers to taking care of a structure and the surrounding property. Repairs help restore a structure to its original condition.

● Alterations are changes to a structure's form. The two types of alterations are restoration and remodeling.

Test Your Knowledge

1. What is the name of the coating used to seal wood before painting?

2. Name the three kinds of paint finishes.

3. What are the two types of bases used in paints?

4. Name three kinds of floor coverings.

5. In landscaping, what is the name given to such things as decks, driveways, and fences?

6. What process gives the earth the shape called for in the landscaping plan?

7. List three kinds of maintenance done in most homes and other buildings.

8. Name the two types of alterations to a structure.

9. What is another term for restoration?

10. What is the term used for returning a structure to its original condition?

Correlations

LANGUAGE ARTS

1. Make a videotape demonstrating the differences among finishing products, such as flat, gloss, and semi-gloss paints. Organize your demonstration carefully. Use visual aids and speak clearly.

SOCIAL STUDIES

1. Write a report on the restoration of the town of Williamsburg or a famous building, such as Thomas Jefferson's home Monticello.

SCIENCE

1. Check with a local nursery to find out which shrubs and trees look best all year round in a cold climate. They should either have foliage all year long or have an attractive shape when they lose their foliage. Create a landscaping design for your school from this information.

MATH

1. Suppose you are buying carpet for your living room, which measures 12' × 15'. The carpet you like is $15.99 per square yard. How much will enough to cover the floor cost?

You · Can ·

— Make a Difference —

Teens Build a Home for Other Teens

What would happen if you had no place to sleep tonight? What if it were snowing outside and you had no shelter? If you lived in the South Bronx, New York, you might call Banana Kelly.

Banana Kelly?

Who's that, Superman in a yellow suit?

Well, super maybe, but that's where the resemblance ends. Banana Kelly is a neighborhood housing group. Along with Bronx Regional High School, Banana Kelly has purchased a five-story abandoned building from the city for a cost of $1. Technology students from the high school are turn-ing the building into a dormitory for homeless students. The New York State Department of Social Services has supplied over one million dol-lars for materials and tools. Other groups and private ind-ividuals are providing money or supplies.

Bronx Regional High School enrolls its technology students in two nine-week courses. During the first nine weeks, each 15-member group learns to use measuring and power tools, recognize building materials, and understand a builder's vocabulary. During the second nine weeks, each team learns other skills that will prepare them for the job market. While they are learning, they are helping create the dorm. The members of the team work hard. They are paid $4.50 an hour and get course credit for the program.

Placement counselors say that about 45 percent of the students are eventually placed in jobs, although not all jobs are in the construction industry.

The students are proud of their accomplishments. When finished, the dormitory will be home to high school students who are either homeless or victims of domestic violence. It will house eight females and twelve males on separate floors. It will have dining facilities, a library, a recreation room, and living units for adult supervisors.

The students who are working to remodel the building know all about homelessness and domestic violence. Many of them have had a hard time making it through traditional high school environments. The building program is helping them turn their lives around. One worker, a former drop-out, now wants to graduate from high school and go on to college. Another feels he is building something for the future. He will be proud to have other people living in the dormitory he helped create.

LOOKING AHEAD

Activity Brief:
Building a Future
PART 1: Here's the Situation...........

During the past 50 years, new inventions and processes have changed the way we live. Fifty years ago, most people thought it was impossible to send a human being to the moon. Now we have not only reached that goal but have sent probes much farther into space.

The year 2000 is not far away. For many people that year is a symbol of the future. With it we will enter the 21st century. The 21st century will bring changes in technology, in the world, and in our personal lives. Will you still be in school then? Perhaps you'll have entered the world of work. What kind of job do you think you will have? Will it be in production systems?

This section is about the future of production systems. It is also about your own future. In Chapter 17 you will learn how the exploration of outer space and other technologies may affect production systems. In Chapter 18 you'll discover some things you can do to make your own future brighter.

PART 2: Your Challenge ....

This activity is one you will design yourself. Suppose you're a famous problem-solver, known all over the world. The leaders of a country that is trying to build a better future have called you in to help them solve a problem. You can solve their problem doing something you do best. What will it be?

Look at Tables A, B, and C. Table A lists possible countries to choose from. Table B lists problems those countries could have. Table C lists ways in which you might present your solution. To create your activity, choose a country from Table A and a problem from Table B. Then consider the projects listed in Table C. You may choose one of the projects from Table C or, with your teacher's approval, you may do something quite different.

Then, when you are finished, give a presentation of your solution to the country's leaders (in this case, your classmates). Sell them on your idea. Tell them why you chose it and why you believe it will help their problem.

Research Tip:
The Readers' Guide to Periodical Literature lists almost all current magazine articles that have been published recently. They are organized by subject.

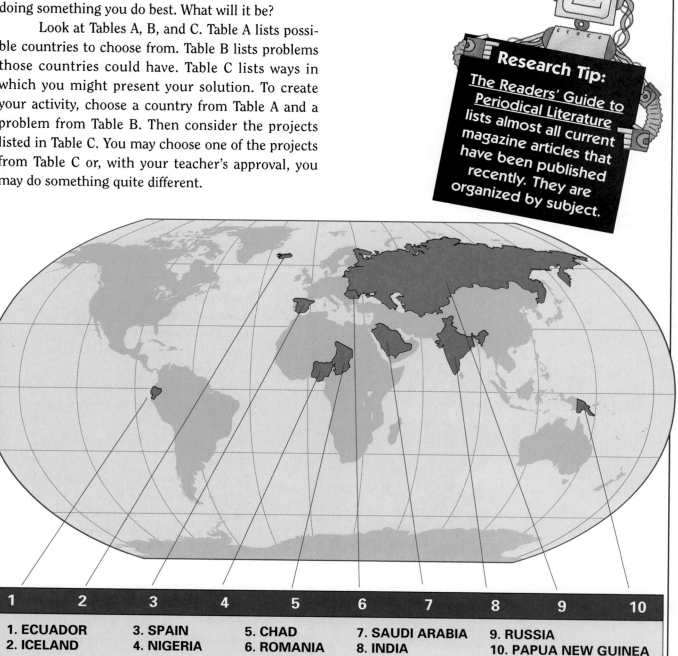

1	2	3	4	5	6	7	8	9	10

1. ECUADOR 　 3. SPAIN 　 5. CHAD 　 7. SAUDI ARABIA 　 9. RUSSIA
2. ICELAND 　 4. NIGERIA 　 6. ROMANIA 　 8. INDIA 　 10. PAPUA NEW GUINEA

Table A. Countries To Choose From

See the map on page 279 for where these countries are located. Read the bits of information included here about each one. Then pick the one that will "hire" you.

Chad	Papua New Guinea
Ecuador	Romania
Iceland	Russia
India	Saudi Arabia
Nigeria	Spain

INDIA

LOCATION: In Asia; the 7th largest country in the world.

LAND: Himalayan Mountains; vast plains and plateaus, jungle.

ECONOMY: Farming and industry; huge population; large cities and many rural areas.

Table B. Problems To Choose From

Choose one of these problems for your country. Pick one in a career area that interests you.

- the country wants to develop a low-cost form of energy production
- the country wants to conserve the energy it already produces
- the country has a water shortage
- the country wants to turn some of its natural resources into a product that can be exported
- the country wants to produce a low-cost, healthy food that will feed many hungry people
- the country needs low-cost housing for its cities
- the country needs low-cost housing for undeveloped areas
- the country wants to build a structure that will help tourists understand its culture
- the country needs a system of roads for its undeveloped areas
- the country needs a bridge design that will handle car traffic as well as more primitive methods of transportation, such as ox-carts

Table C.
Some Methods of Presentation

These are only some of the methods you might use to show how you would solve the problem. You need only pick one. However, you may do more than one if you wish.

- design and build a product or device
- make several kinds of drawings
- make a model or prototype
- create a videotape

SPAIN

LOCATION: In western Europe.

LAND: High plateaus and mountains; sunny climate.

ECONOMY: Based on farming, industry, and tourism; few natural resources; half of the people live in cities.

PART 3: Specifications and Limits...........

Your project will need to meet certain standards. Read the following specifications and limits carefully before you begin.

1. Your solution must be appropriate to the country you've chosen. For example, if you choose food production in a desert country, you will have to take the lack of water into consideration.

2. Your solution must not create obvious negative impacts that destroy its value. For example, suppose you choose to build a factory in a country with few natural resources. Where will the factory get the materials? If it's a poor country, buy-

ing the materials elsewhere will only make it poorer. This does not mean you shouldn't build a factory there. You must simply try to see the whole picture.

3. You must hand in the following:
   - a daily log sheet of your work
   - a statement identifying the country and problem you chose
   - all sketches, drawings, posters, or storyboards
   - any models, prototypes, videotapes, or devices you've made
   - all your research notes
   - a paragraph describing your solution and why you chose it
   - a list of impacts, both positive and negative, that your solution could have on the country, the people, and the rest of the world
   - a list of resources you used, including books, magazines, interviews, etc.

4. You will be evaluated on the basis of:
   - originality
   - how well you researched the problem
   - how usable your solution would be
   - how well you took different conditions into account

PAPUA NEW GUINEA

LOCATION: A series of islands in the Pacific Ocean north of Australia.

LAND: Mountains, jungle, and swamplands.

ECONOMY: Based on farming; more than 700 languages spoken.

ECUADOR

LOCATION: One of the smallest countries in South America.

LAND: Andes Mountains; valleys, flat jungle areas.

ECONOMY: Based on farming; a fast-growing population.

PART **4**: Materials...........

There are many materials you might use. Here is a list of possibilities.

Materials

paper and pencils
drafting tools and
   equipment
computer and CAD
   software
cardboard, plastic, wood,
   sheet metal, foam core
   board, posterboard,
   rubber bands, dowels,
   and other materials
   generally available in
   the lab
video camera and tapes

PART 5: Procedures..........

The problem you choose and how you solve it will be up to you. Still, there are certain steps to follow that will make your work easier.

1. Be sure to keep a log of your work each day.

2. Look at the list of problems in Table B and pick one that interests you. Think of it as a "career" problem. In other words, solving it should involve you in work you think you might want to do someday. For example, if engineering interests you, you might choose water shortages as your problem. Then you could design a dam or pipeline to solve the problem.

3. On these pages you'll find bits of information about the countries listed in Table A. Pick one that interests you.

4. Look at Table C. On it are ways in which you might present your solution. If you wish to choose a different method, first get your teacher's approval.

5. Inform your teacher of the choices you made from Tables A, B, and C.

6. Visit the library. Do some research on the country you chose. Where is it located on the map? What is the climate like? The terrain? What resources does it have? Do its people live in cities or small villages? Are they rich or poor? Do they farm the land or work in factories? This information will help you learn what's possible.

7. Now think of the problem in terms of the country. What does the country have available that might help you solve the problem? What must it obtain that it does not already possess?

8. Try several solutions. Write your ideas down or make sketches. Choose the one you like best. Submit it to your teacher for approval.

9. Put your idea into final form.

10. Test your idea by going back over your research. Will it really solve the problem? If not, make adjustments.

CHAD

LOCATION: Fifth largest country in Africa.
LAND: A level plain with grasslands and desert; one large lake.
ECONOMY: Poor; based on farming; little industry, small population.

ICELAND

LOCATION: An island in the North Atlantic below the Arctic circle.
LAND: Large glaciers, steaming hot springs, volcanoes.
ECONOMY: Based on fishing and fish processing.

PART **6**: For Additional Help...........

The help you need will depend on the country and problem you choose. You should be able to find enough information about the country in an ordinary encyclopedia. As for the problem, if you've chosen one that interests you, you probably already have an idea of what to research. If not, start with the general subject and work toward specific things.

For example, suppose your country wants to develop low-cost energy. First, look up energy in this text and in the encyclopedia. Find out what types cost the least. Which of those would work in your country? Solar energy is low-cost. If your country gets a lot of sunshine all during the year, your solution might include its use. If it is a northern country with little sun, you'll have to look further.

Next, you might want to come up with a solar device the country could produce. For that, you would first find out more about solar energy. Ask your librarian for sources. The more you explore, the more help you'll find.

ROMANIA

LOCATION: In eastern Europe; formerly a member of the Soviet Union.

LAND: Mountains, forests, sandy beaches on the Black Sea.

ECONOMY: Many farms, factories, power plants, and resort areas; formerly communist, now trying free enterprise.

RUSSIA

LOCATION: North of Europe; the largest country in the world; formerly in the Soviet Union.

LAND: Grassy plains, thick forests, high plateaus and mountains.

ECONOMY: Based on farming and industry; formerly communist, now trying free enterprise.

PART 7: How Well Did You Meet the Challenge?...........

When you've finished your solution, evaluate it. Does it do what it's supposed to do? Ask yourself the following questions. Your teacher and classmates may take part in this evaluation.

1. How well does your solution solve the problem you chose?

2. How well does your solution take into account the conditions of the country?

3. Did you do enough research on your subject?

NIGERIA

LOCATION: On the west coast of Africa.

LAND: Hot, rainy swamplands; dry, grassy plains; tropical forests; high plateaus and mountains.

ECONOMY: Based on farming and fishing; many mineral resources; more people than any other African country.

SAUDI ARABIA

LOCATION: To the east of Africa.

LAND: Vast deserts and some fertile areas; mountains.

ECONOMY: Most people are farmers; huge deposits of petroleum make the country a leading economic power.

PART 8: Extending Your Experience...........

Think about the following questions and discuss them in class.

1. What conclusions can you draw from this activity about solving problems?

2. Name one new thing you learned during this activity.

3. How do you think you would like working in the country you chose? Why?

4. If you could do this activity again, what would you do differently?

5. Who in real life is doing work of this kind?

CHAPTER 17

A Look at the Future

Introduction.......................................

When you're an adult with a job and family, what will the world be like? One thing is certain: it will be different than it is today. To begin with, production systems will have changed. New methods and materials will be used in manufacturing and construction. New products will change the way you work, play, and eat. In this chapter, we will consider some of these changes that are coming.

After studying this chapter, you should be able to......................

Discuss the techniques futurists use to make predictions.

Describe new technologies that have resulted from the space program.

Discuss current technologies used in biotechnical production.

Tell what the future has in store for manufacturing and construction.

Words you will need ....

airspace construction
artificial intelligence
biotechnology
Canadarm
Delphi Study
futurists
genetic engineering

lean production
microgravity
simulation
smart buildings
spinoffs
tissue culture
trend analysis

Predicting the Future

People who try to predict the future are called **futurists**. Futurists work in many fields, such as medicine and technology. They analyze what has happened in the past. They also observe what is happening now. They pay close attention to research going on in different fields. Then they put all this information together and try to guess where it will lead. The methods used by futurists include trend analysis, surveys, scenarios, and simulations.

> ## IMPACT
>
> **Because the construction industry is sensitive to changes in the economy, futurists look at new construction activity to help them make predictions. The number of new building permits issued is one of the facts considered. It helps indicate the general direction — up or down — of the overall economy.**

Trend Analysis

A trend is a tendency for large numbers of people to do something. For example, a current trend is for people to eat healthier foods. In other words, many people are beginning to do it. **Trend analysis** is the study of trends in order to identify important ones. Futurists use trend analysis to make predictions.

They review newspapers, magazines, and research reports to learn what topics are being written about frequently. Fig. 17-1. Currently, they are finding many articles on the environment and on new uses for computers. They analyze these articles and do some of their own research.

Fig. 17-1. What trends do these pictures and advertisements show?

Surveys

Surveys are also helpful to futurists. Futurists ask people a series of questions and compare the answers. The answers help determine how people feel about something. Often futurists survey experts. For example, they may want to know what kind of management styles will be popular in the future. The experts they survey might include current company presidents, managers, and business teachers.

One such survey of experts is called the Delphi Study. The panel of experts for the **Delphi Study** works together to predict the future. A question is asked and each panel member submits an answer in writing. Each member receives a copy of the answers given by each of the experts. Everyone then prepares another written answer. This process continues until the panel finally reaches some agreement.

Scenarios

Scenarios are like stories. They describe possible situations. Futurists use scenarios to help people think about important issues. For example, some people are very concerned about the pollution caused by fossil fuels, such as coal, oil, or gas. They think it causes the earth to warm up too much. Jeremy Rifkin, the author of several books, created this scenario about the effects of global warming:

The year is 2035. In New York, palm trees line the Hudson River from 125th Street to the Midtown exit.

Phoenix is in its third week of temperatures over 170 degrees, and the project to cover the city with air-conditioned domes is still unfinished.

Not everyone believes that Rifkin's scenario presents an accurate picture. However, it does get people's attention.

Simulations

Futurists use simulations to test their ideas. A **simulation** is an imitation of something. Pilots, for example, learn how to fly with the help of a flight simulator. The simulator has a cockpit and its windshield shows them a make-believe situation.

Futurists often use computers to create simulations of possible events. They ask the computer, "What will happen if…?" For example, scientists have asked computers to help predict the effect of a nuclear war on the world's climate. The computer then suggests the different things that might happen.

▶▶▶ **FOR DISCUSSION** ◀◀◀

1. Why do you think our government is interested in what futurists have to say?
2. Based on your own reading and observations, describe three major trends going on today.

Extension
Activity

■ Write a 150-word scenario describing a house being built in the year 2020.

Using Space Technology

When people think about the future, the exploration of space usually has a place in it. Fig. 17-2 (pp. 290-291) shows some high points of space exploration. Since its founding in 1958, the American space program has helped bring about many advances in technology. The space shuttle has become the latest means for increasing our knowledge.

Materials Processing

There is no air or gravity in space. Certain experiments can be done there that can't be done on earth. As a result, the space shuttle is being used for research projects to

- manufacture products that cannot be made very well on earth, Fig. 17-3
- learn how to improve materials processing on earth
- make new discoveries that may be useful commercially.

Although there is no gravity in space, the shuttle is somewhat affected by the earth's gravity. Because the gravity is so low, it is called **microgravity**. Where there is microgravity, materials of different densities can be mixed. For example, have you ever tried to mix oil and water? On earth it can't be done. Because they are of different densities, they separate. The denser material (water) settles to the bottom. In space, new alloys can be made because this settling does not occur. Many other types of products might also be produced in an environment with microgravity. They include crystals for use in electronics, new medicines, ceramics, and glass.

NASA wants this new knowledge to be useful for all of us. To encourage this, it has set up a series of Centers for Commercial Development of Space. Each center specializes in a different aspect of space technology.

10-Micrometer Polystyrene Sphere

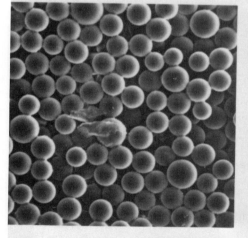

Earth-Made Particles

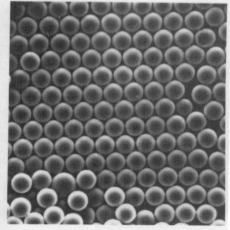

Particles Made in Space

Fig. 17-3. These tiny spheres are used to adjust electron microscopes. The spheres made in space (right) are much more uniform in size than those made on earth (left).

October, 1957: The first human-made satellite, *Sputnik I,* is launched from the Soviet Union.

July, 1958: NASA is founded to plan and operate the American space program. It immediately launches a series of satellites to photograph the earth and moon.

September, 1959: The Soviet Union launches the first probe to strike the moon.

Fall, 1959: NASA conducts the first test of a Mercury capsule designed to carry people into space.

April, 1961: Russian cosmonaut Yuri Gagarin becomes the first human being to orbit the earth.

1961: President John F. Kennedy challenges NASA to put an American on the moon by the end of the decade.

May, 1961: During a 15-minute *Mercury 3* flight, Alan Shepard becomes the first American in space.

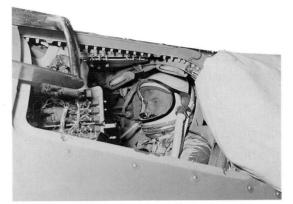

February, 1962: American John Glenn pilots the *Mercury 6* capsule in an orbit around the earth.

1965-1966: Twelve Gemini spacecraft are launched. Their goal is to prove that humans can live and work in space.

June, 1965: Ed White is first American to walk in space.

1967: During a ground test, a fire breaks out in the *Apollo 1* capsule and three American astronauts are killed.

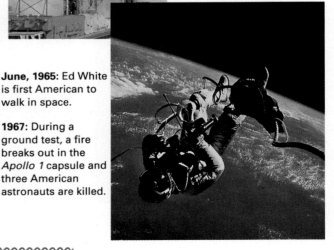

Fig. 17-2. This time line shows some of the history of space exploration.

1968: Three American astronauts orbit the moon in *Apollo 8.*

May, 1973: The first American space station, *Skylab 1* is launched. For six months it is used by three different crews of astronauts to conduct laboratory experiments.

January, 1972: President Richard M. Nixon approves development of the space shuttle.

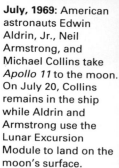

July, 1969: American astronauts Edwin Aldrin, Jr., Neil Armstrong, and Michael Collins take *Apollo 11* to the moon. On July 20, Collins remains in the ship while Aldrin and Armstrong use the Lunar Excursion Module to land on the moon's surface.

1969-1971: During the Apollo program, twelve different American astronauts land on the moon. Samples of lunar rock are brought back for study.

July, 1975: For the first time, Russian and American spacecraft rendezvous, and the astronauts and cosmonauts work together.

July 1971: The lunar rover becomes the first motorized vehicle used on the moon.

1971: The Soviet Union establishes the first space station, *Salyut.*

April, 1981: The first American space shuttle, the *Columbia,* is launched from the Kennedy Space Center at Cape Canaveral, Florida. Since then, four shuttles have been built: *Challenger, Discovery, Atlantis, and Endeavor.*

Fig. 17-4. The array of solar panels (left) generates energy in the Mojave Desert, California. The panels on top of the house (above) provide heat and hot water. The panel at the top of the calculator (right) collects solar energy for power.

Spinoffs

Ever since it was founded in 1958, NASA has been the world's leading source of new technologies. The products and processes resulting from these technologies are called **spinoffs**. To date, there have been more than 30,000 such spinoffs. Areas in which these spinoffs have been developed include energy, the environment, medicine and health, production, and consumer goods.

Energy

Energy is needed for space travel. Since satellites have limited storage and cannot easily return to earth for fuel, they must produce their own energy. They must also use that energy efficiently. Research into these needs has helped solve some energy problems on earth.

When silicon is exposed to sunlight, its electrons are set into motion. These moving electrons can be taken from the silicon in the form of electricity. This is called photovoltaic conversion. NASA uses photovoltaic conversion to provide electricity to satellites.

Photovoltaic, or solar, "cells" are made by electrically connecting very thin slices of silicon. The cells are then packaged in plastic or glass for protection. Several cells can be connected to form panels. When many panels are connected, the arrangement is called an array. These arrays are quite large.

Photovoltaic power is being put to many uses on earth. Fig. 17-4. It provides electricity for villages in remote locations. It operates railroad signals, navigation buoys, and emergency telephones along highways. Even small, pocket calculators are made to work with solar power. Arco Solar, an American company, is installing photovoltaic power plants all over the world.

The Environment

Water must be provided for astronauts during long space flights. Since storage space is limited, waste water must be reclaimed during the voyage. To solve the problem, NASA scientists used a natural water purification system. They discovered that the water hyacinth, a free-floating fresh water plant, can digest large amounts of pollutants. Today, these and other plants are being used for sewage treatment in the southern United States. Research on plants that will do the same thing in colder climates is under way.

Similar research has been done to learn how to purify the air in space capsules. Scientists discovered that common houseplants absorb many pollutants. Now several companies are trying to develop new filtering systems that include natural methods.

IMPACT

Because construction uses up available land, some people feel certain buildings are wasteful. For example, less land would be needed if everyone lived in high-rise apartment buildings. Some designers believe the city of the future should be one enormous building surrounded by many acres of open land.

Medicine and Health

During missions in space, NASA keeps a close check on the health of astronauts. Some of the technologies developed are being used to keep people on earth healthier. Tiny portable heart monitors used during astronaut training are now being worn by heart patients. The monitors keep track of heart activity as the people go about their normal lives.

Robotic technologies developed for use in space have been applied to wheelchairs. The wheelchairs can be operated by voice commands. A computer built into the chair recognizes the user's voice. It responds to commands such as "go," "forward," and "stop."

Production

Practical spinoffs have also resulted from the fifty-foot robotic arm used on the space shuttle. It is called the **Canadarm** and is used to launch and repair satellites. Fig. 17-5. The arm was designed by Spar Aerospace of Toronto, Canada.

Fig. 17-5. This robot arm helps astronauts work in space. The curve of blue in the picture is the earth.

A similar robot is now being used for dangerous jobs in nuclear power plants. The arm is remote-controlled from a safe location somewhere in the plant. A similar remote-controlled arm is being used to attach wire screening to the roof of mines. The screening makes the mines safer to work in.

Consumer Goods

Many products that we use today were originally developed for the space program. For instance, do you use a pocket calculator or a digital watch? Both resulted from the need to make things smaller for use in early spacecraft. Do you own a pair of sneakers or a jacket that closes with Velcro™? Velcro™ was first used during Mercury flights to keep items from floating around the capsule.

Fire within a space capsule is especially dangerous. Flame-resistant materials developed to reduce fire hazards are being used for seat cushions in airplanes. Because smoke and fire cause many of the deaths following plane crashes, these materials can save lives. Fig. 17-6.

Because cans of food are too heavy and hard to handle in space, special containers had to be developed. The result was a lightweight and strong container called a retort pouch. Retort pouches are made of aluminum foil sandwiched between layers of plastic. To heat the food, the pouch is simply placed in boiling water.

Many skiers now wear electrically heated goggles. The goggles are a spinoff of the system used to keep astronauts' helmets fog-free.

NASA is now working on new technologies for the space station *Freedom*. Several projects are expected to produce spinoffs. They include:

- **Artificial intelligence.** Artificial intelligence is the term used for certain advanced computer programs. The programs enable computers to learn from experience. They then apply that knowledge to new situations, just as humans do. Using artificial intelligence, computers will make faster and more accurate decisions than humans can.
- Advanced space suits. These new space suits will be safer and more comfortable. They will be used for building and maintaining the space station.
- Growing food in space. NASA is experimenting with crops to learn which ones will grow best in space. It also hopes to increase the amount of food that can be grown per square foot of growing space.

Fig. 17-6. Flame resistant materials make airplanes safer.

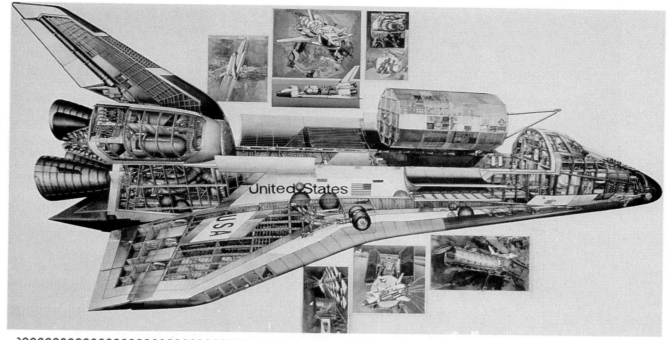

Fig. 17-7. This is a cutaway view of the Space Shuttle. Which areas can you identify?

The Space Shuttle

Today, the space shuttle is the most important part of NASA's Space Transportation System. Fig. 17-7. The shuttle is unique. It is launched like a rocket but lands like an airplane. The orbiter can carry a crew of two to seven. It circles the earth in a low orbit. The rocket boosters and orbiter are reusable.

The shuttle can be used for many things. It can be used to launch and repair satellites and carry cargo to and from space. It can serve as a laboratory for science and technology experiments. In the future, the shuttle will be used to carry the workers, materials, and equipment used in building the new space station. NASA also plans to develop more unmanned vehicles for launching satellites. The shuttle will then be used only for missions that need a crew.

The Space Station

The space station *Freedom* should be in operation by the year 2000. It is an international project involving NASA, Canada, Japan, and the European Space Agency.

Freedom will serve as a base for exploring the solar system. It will also be used to continue microgravity research and materials processing. Medical experiments will study how humans adapt to life in space. This information is needed to learn more about the effects of long space voyages.

The station will orbit 250 miles above the earth. It will get electrical power from four solar arrays located at the ends of a 495-foot-long truss. At the center of the truss will be the living quarters and laboratories. Fig. 17-8.

Fig. 17-8. Six to eight people will live and work in the space station all year long. NASA expects the station will be useful for as long as 30 years.

A crew of six to eight people will operate the station 24 hours a day, 365 days a year. The shuttle will carry crew members and supplies between the station and earth about every three months.

NASA expects that knowledge gained at the station will make manned flights to Mars and other planets possible.

The Pathfinder Program

Pathfinder is a NASA program that will develop new technologies for exploring space. One project will be to design a permanent moon base. Another will be to develop new vehicles and ways to operate them. These systems would be used during robotic and manned explorations of Mars. The program will also devise techniques needed to assemble spacecraft in outer space.

▶▶▶ FOR DISCUSSION ◀◀◀

1. How have satellites changed communications? How has your life been affected?
2. Why do you think NASA wants to grow food in space? What do you think the difficulties would be? How do you think the knowledge gained might be used here on earth?

The Future of Production Systems

The future holds many things in store for production systems. Futurists think that energy, biotechnical production, manufacturing, and construction will all be affected.

Energy

The cost of energy will continue to rise. This will add to the cost of production. Alternate energy sources, such as wind and solar power, will have to be developed further.

Many experts recommend using more natural gas instead of oil in the future. Conserving energy will continue to be important. More efficient heating and cooling systems will have to be devised. Buildings will be designed so they can take advantage of the sun's energy. The cost of photovoltaic cells is also expected to drop sharply. These cells will be used in addition to traditional energy sources.

Manufacturers will continue to create better insulating materials. For example, window films that let in the sun's heat during the winter and keep it out during the summer should have widespread use. New building methods will also help reduce energy use. Walls made using 2" x 6" studs may become standard in residential construction. Extra insulation can then be placed inside the walls for greater comfort and energy savings.

Biotechnical Production

Biotechnology is based on the use of living things, including plants, animals, and microorganisms. (Microorganisms are plants or animals too small to see with the naked eye.) Biotechnology is an exciting field that will bring great changes to the production systems of the future. It includes genetic engineering, plant tissue culture, controlled environment agriculture, and aquaculture.

Genetic Engineering

Genetic engineering is the science and technology to alter DNA. DNA (deoxyribonucleic acid) is the basic building block of life. Genes, which determine the characteristics of living things, are made of

DNA. By changing the genes, it is possible to create a virus, plant, or animal not found in nature. Fig. 17-9.

Genetically engineered farm animals serve as factories to produce rare medicines used to treat human diseases. These medicines include human insulin used by diabetics, a human growth hormone used for children with growth problems, and a substance to help blood clot in people with hemophilia (uncontrolled bleeding).

Some people are against genetic engineering. They do not approve of experiments that may cause animals to suffer. They do not believe that animals should be used like machines. There is also concern about releasing new forms of life into the environment. How will the new life forms interact with other life forms? Will they create problems worse than the ones they were designed to correct? These and other questions must be carefully considered.

Tissue Culture

By using **tissue culture**, whole plants are grown from tiny pieces of the parent plant. The new plant is called a clone. A clone is an exact genetic duplicate of the parent.

Tissue culture is becoming very useful in agriculture. Most houseplants and many crops can be produced by using tissue culture. Researchers hope to develop new plants that are more resistant to drought and disease.

Controlled Environment Agriculture

Controlled environment agriculture is done in large greenhouses. Computers control temperature, light, humidity, and nutrition. In the future, controlled environment agriculture will allow food crops to be grown year-round in all climates.

Fig. 17-9. Genetically engineered mice bred by researchers at Harvard University are being used for cancer research.

Fig. 17-10. These tiny plants are being grown without soil.

Hydroponics is one type of this agriculture. In hydroponics, plants are grown without soil. Instead, sand, rock-wool insulation, or water is used. Fig. 17-10. Liquid fertilizers supply nutrition. Hydroponics conserves water and allows for better control of plant nutrition. Plants grown this way also get fewer diseases.

Aquaculture

Aquaculture is the controlled production of plants and animals that live in water. Most aquaculture systems today are fish farms. About 10 percent of the seafood we eat comes from aquaculture. As more fish farms are built, this percentage is expected to increase.

Manufacturing

What do futurists predict for manufacturing? They say that the future will bring many changes. New materials and methods will continue to be developed. Others may disappear.

Because many natural resources are being used up, alternatives must be found. For example, some experts say we have enough oil for only 30 to 50 more years at the current rate of use. This means plastic will have to be made from a different material. Also, cars will have to use different fuels. Fortunately, recycling of materials will continue to grow. More and more products will be made from recycled materials.

Fig. **17-11.** Car bodies are welded on this assembly line. The process is automated and requires no workers.

Fig. **17-12.** Accountants are being retrained by this company for new jobs.

Quality assurance will become more important. Today, it is the function of the engineering or quality control departments. In the future, every department in a company will play a part in product quality.

More and more production processes will become automated. Costs will be lower and production will be faster. In the distant future, some factories may have no workers at all. Fig. 17-11. Computers will do everything, including designing the machines used to make products.

Fewer human workers will be needed to manufacture most products. Many kinds of jobs will no longer exist. New jobs will be created by the new technologies. Workers will have to be retrained to do such things as maintain and repair computers and robots. Continuing education will become necessary for most people. Fig. 17-12. Because change will come so fast, many workers may have to learn several jobs in their lifetimes.

Lean Production

A new production method used in Japan is expected to spread around the world. It is called **lean production**. Lean production requires less labor, less engineering time, and less inventory than regular manufacturing. Experts predict it will become the standard production method in the 21st century.

In lean production, teamwork is the key. In fact, workers are organized into teams. Each team has a leader. The team determines the best way to do the job it is assigned. It is also responsible for making sure the job is done right. Fig. 17-13.

Teamwork has helped to improve product quality. Each team member can stop production if he or she sees a problem. Teams are also given time to work on improvements in the production process. Each worker has greater pride in a job well done. The employees feel they are important to the company.

Fig. 17-13. This automotive production team is trying to figure out a better way to do a job.

Fig. 17-14. A pneumatic nailer can drive nails many times faster than a human worker with a hammer.

Workers are not limited to one job. They are encouraged to learn new jobs at the plant. This helps give them a better sense of how the company operates. Work is more challenging and less boring.

In Japan, suppliers are also part of the production team. They are involved in the design and engineering process. As a result, the quality of products is improved. Deliveries by suppliers are made using the just-in-time system. Inventories are eliminated.

Construction

In 1990, the world population was 5.2 billion. In the year 2000, it is expected to be 6.2 billion. Where will these people live and work? Futurists predict that changes in construction technology will help meet their needs.

Combining Manufacturing and Construction

In the future, construction will continue to use manufactured parts in buildings. Almost all buildings will have factory-made roof trusses. The trusses will reduce design and construction costs.

The use of specialized equipment will increase. Laser surveying equipment will replace the equipment now in use. Laser levels will help carpenters create interior walls with more accuracy. Many carpenters will use pneumatic nailers instead of hammers. Fig. 17-14.

Panelized, sectional, and modular construction will become more important. In panelized construction, floor, wall, and roof panels are built in factories. After the panels are assembled at the site, the rest of the structure is finished.

In sectional construction, the complete structure is made at the factory. A three-bedroom, ranch-style house can be made in two sections and carried to the site by truck. A crane places the sections on the foundation. Fig. 17-15. Combining the two sections into a finished house takes only a few days.

Modular construction has been used successfully in building motels. Identical units are delivered to the site and assembled. Each unit comes with plumbing, electricity, heat, and air conditioning already installed.

Smart Buildings

By the year 2000, most new houses will have computer management systems. Several hundred of these **smart buildings** have already been built around the United States. Computers control their day-to-day operations, including utilities, security systems, and appliances. Fig. 17-16.

Fig. 17-15. This house was completely built at the factory. It needs only finishing touches.

Fig. 17-16. Computers control all major systems in smart buildings.

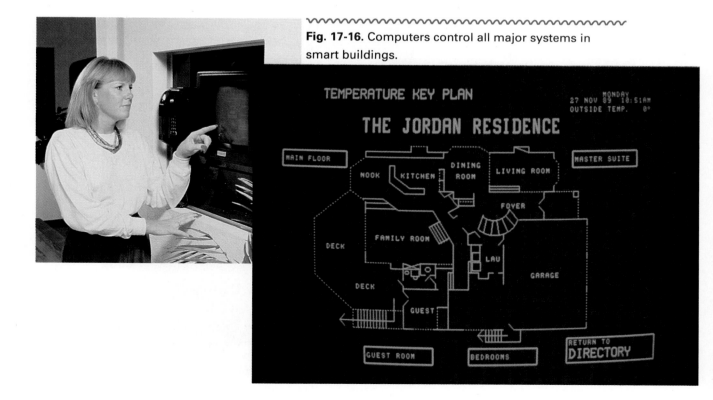

Fig. 17-17. In the future, people may live in underground cities like this one. What would be the advantages to living here? The disadvantages?

These systems help conserve energy. When a person enters a room, devices called sensors tell the computer to turn the lights on. When the person leaves, the computer turns the lights off. Heating and air conditioning are also computer controlled. Appliances such as dishwashers operate when utility rates are lowest. Security systems warn of smoke, fire, or intruders. If needed, the police or fire department is automatically notified.

New Construction Sites

Today, most structures are built on available land sites. However, in many cities undeveloped land is in short supply. New sites will have to be found.

Underground cities. One solution to the shortage of sites may be to build underground. In Crystal City, a project in Arlington, Virginia, most residents live above ground in high-rise buildings. Movie theatres, shops, restaurants, and other facilities are below ground. Fig. 17-17.

Japan is a small country with a large population. There is a shortage of open land. Several construction companies there are planning underground cities. Some of these cities will be new. Others will be built below existing cities.

Airspace construction. When a structure is built above a highway, it is called **airspace construction**. Fifty airspace projects have been completed in California. Several hundred others are planned. The projects include apartments, restaurants, parking areas, and highway rest areas. Fig 17-18.

Undersea construction. Water covers three-fourths of the earth. Scientists are just beginning to learn about what goes on in our oceans. In the future, large modules will be transported by ship and assembled at sea. These modules will form self-contained living and working quarters for teams of scientists. The scientists will be looking for deposits of

valuable minerals. They will also conduct research to develop new sources of food and medicine. One day, large undersea cities may become practical.

▶▶▶ FOR DISCUSSION ◀◀◀

1. **What energy conservation measures have been used in schools, stores, and other public buildings in your town?**
2. **Think about the pros and cons of genetic engineering. On the whole, do you think genetic engineering is good or bad?**
3. **Suppose you had the opportunity to work in a plant that uses both traditional assembly lines and team assembly lines. If you had the choice, which line would you work on? Give your reasons.**

Extension Activity

■ **Build a model "smart building." Use electronic components — a photovoltaic cell, a battery, a relay, and a bulb — to assemble a system that turns on a light when someone enters a room. The system should also turn off the light when the person leaves.**

Fig. 17-18. The AT & T Gateway Tower in Seattle is built over a highway.

Chapter Highlights

● People who study and predict the future are called futurists. The methods used by futurists include trend analysis, surveys, scenarios, and simulations.

● Products and processes resulting from space technologies are called spinoffs. Areas of development for spinoffs include energy, the environment, medicine and health, production, and consumer goods. Important projects in our space program include the space shuttle, the space station, and the Pathfinder program.

● Biotechnology is based on the use of living things. An important part of biotechnology is genetic engineering, which combines science and technology to alter DNA.

● Lean production emphasizes teamwork. It reduces costs and saves time.

● Smart buildings use computers to control their day-to-day operations.

Test Your Knowledge

1. Name three methods futurists use to predict the future.

2. What kind of survey uses a panel of experts to predict the future?

3. What type of gravity exists on the space shuttle?

4. What are new products and processes resulting from space technology called?

5. By what other name are solar cells known?

6. Which technology is based on the use of living things?

7. What branch of biotechnology combines science and technology to alter DNA?

8. What is the name for controlled production of plants and animals that live in water?

9. How are workers organized for lean production?

10. List three kinds of construction using manufactured parts that will be widely used in the future.

Correlations

LANGUAGE ARTS

1. Using a dictionary, define each underlined part in this list of words. (Notice how each underlined part affects the definition of the entire word.)

<u>micro</u>gravity	<u>hemo</u>philiac
<u>digit</u>al	<u>hydro</u>ponics
<u>uni</u>que	<u>aqua</u>culture
<u>gen</u>etic	<u>bio</u>technology

SOCIAL STUDIES

1. As a class, debate this statement: Production systems can/cannot help bring greater harmony to the world in the future.

SCIENCE

1. Put a short piece of copper pipe on a balance to measure its mass in grams. Then put it into a graduated cylinder that contains enough water to cover the pipe. Record the amount the water rises. This is the volume of the copper in the pipe. Divide the mass by the volume to calculate the density of the copper.

MATH

1. A new motel is being built using modular construction. The four story structure has 23 units on each floor. How many total units will be in the new motel?

CHAPTER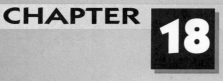

Your Future

Introduction......................................

You have learned a lot in this course about the future of production systems. What about your own future? If you have not already thought about a career, now is a good time to begin. Some things you can do now will help you when you enter the world of work.

After studying this chapter, you should be able to......................

Discuss the personal qualities employers look for when hiring a new employee.

Describe several kinds of schools available that will help prepare you for a job.

List several of your interests, abilities, aptitudes, and values.

Discuss several ways in which you can explore career interests.

Words you will need ....

abilities

aptitudes

attitude

interests

negative attitude

occupational training
  centers

positive attitude

values

vocational training
  centers

volunteer

What Do Employers Look For?

In order to prepare yourself for a career, it helps to know what employers look for. Of course, they want someone who has the right job skills. Those vary with the job. A painter, for example, must know about different types of paint and how to apply them. However, there are general skills useful in every career. These are personal skills that have to do with a person's character and personality.

A Positive Attitude

Attitude is probably more important to success on the job than anything else. Your **attitude** is the way you look at the world, at other people, and at yourself. As a result, it determines how you behave.

An attitude can be positive or negative. People with a **negative attitude** are seldom happy. They do not seem to like themselves or other people very much. They often find fault with themselves or with others and spend a lot of time complaining. Fig. 18-1.

People with a **positive attitude** realize that things can go wrong. However, they try to make the best of any situation. They assume things will improve and don't dwell on the negative part. They figure that things will get better.

Why do employers care about your attitude? A negative attitude makes you hard to get along with. It is the main reason young people lose jobs. They may fight with co-workers or be rude or unhelpful to customers.

A negative attitude may also mean that a person doesn't care about doing a good job. He or she may think, "What's the use?" or "My boss is always picking on me, so why should I try?" Unfortunately, a negative attitude doesn't make the situation any better. It usually makes it worse.

Fortunately, there are several things you can do to help change a negative attitude. Fig. 18-2.

Fig. 18-1. People with a negative attitude can turn others off.

● **You can believe in yourself. Believe that you can learn, change, and succeed.**

● **You can avoid being overly critical of everything. No job, no school, no person is** perfect. You don't have to be perfect either. Lighten up a little.

● **You can give other people a chance. Don't automatically assume that their intentions are bad. They may have a good reason for doing something you don't like.**

● **You can be helpful. Other people will like you for it, and you'll feel good about yourself afterward.**

● **You can accept your own mistakes. Everyone makes them, and your life won't be over when you make them too. Most importantly, don't blame them on someone else.**

● **You can try to find *something* positive about every situation. It's the best exercise for a positive attitude.**

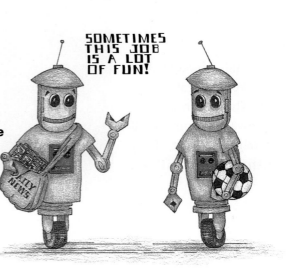

Fig. 18-2. These are keys to a positive attitude.

Willingness to Learn

Do you like to try new things? Are you curious about how things work? If so, you are willing to learn. Because there are many things to learn about any job, the willingness to learn is important. Those who have it are often given increasing responsibility.

Ability to Follow Directions

Most jobs have to be done a certain way. There is usually a good reason for this. Suppose you've been given the job to paint a floor. Your boss tells you to use oil-base paint. You can't find the oil-base paint so you use water-base paint instead. Later, your boss tells you that a special sealer must be put on the floor. It will not stick to water-base paint! You have to remove all the paint and start over. This results in more work for you and costs your boss money and time. This is why employers appreciate workers who follow directions.

Ability to Solve Problems

Sometimes a worker has no directions to follow. She or he will have to solve a problem independently. Employers like workers who can find new solutions to problems on the job.

Willingness to Cooperate

Many jobs today require people to work in teams. In the future, teamwork will be emphasized even more. The ability to work well with others is important to a team's success. Fig. 18-3.

Fig. 18-3. People must cooperate in order to get the job done.

Dependability

People who are dependable show up for work every day and are on time. They know what has to be done and stay until the job is finished. If workers aren't dependable, the work will not get done. Others will have to do it. Deadlines may be missed, and there will be additional expenses. Employers want people they can rely on.

Honesty

Honesty can mean many things. It can mean admitting you have made a mistake and asking for help. It can mean giving your employer a full day's work for a day's pay. It can mean not stealing from the company. Employers invest time and money in workers. They want those who are honest and trustworthy.

▶▶▶ FOR DISCUSSION ◀◀◀

1. Have you ever worked with people who did not have some of the qualities described in this section? How did you feel about them?
2. What other qualities might make someone successful in a career? Why?

Company Activity

■ **Visit the personnel department of a local manufacturing or construction company. Ask a company representative about the personal traits the company looks for when hiring new workers. Ask about the qualifications for a worker who does the jobs you did for your student company. Report your findings to the class.**

How Should You Prepare?

Today, a good education is more important than ever in finding a job. Most jobs in manufacturing and construction require at least a high school diploma. Education can also help you find the career you would be happiest and most successful in. It does this by providing you with information. You learn about the world around you. Fig. 18-4.

IMPACT

The growth of a global economy has changed the way manufacturers do business. Many are seeking markets in other countries. This may change the requirements for some job applicants. Knowledge of a foreign language may be a key to getting some jobs in the future.

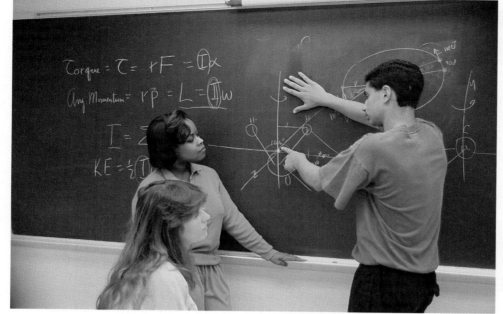

Fig. 18-4. Everything you learn contributes in some way to your future.

General Education

A general education consists of courses such as English, mathematics, social studies, and science. The basic skills you learn in these courses are like a foundation. They give you something to build on.

Some students wonder if such skills are always necessary. They may say, "I only want to work on an assembly line. Why do I need to study English?" There are many reasons. Let's look at a few.

First, workers often have to read instructions to learn how to do something. They must understand the ideas as well as the words they read. If they can't read very well, they may miss something important.

Second, all jobs require the ability to communicate. Good grammar helps people communicate well, both when speaking and writing. If workers can't write a sentence that says what they mean, they will be misunderstood. Fig 18-5. This may hurt their job performance.

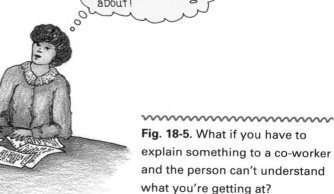

Fig. 18-5. What if you have to explain something to a co-worker and the person can't understand what you're getting at?

Fig. 18-6. These students have graduated from high school. They are now learning carpentry at a vocational training center.

Third, today's jobs are changing rapidly. Most workers will have to be retrained at some point in their lives. The better their basic skills are, the easier it will be for them to learn new tasks later.

The same kinds of reasons apply to math, science, and social studies skills. The more you know, the more you can do. The more you can do, the easier it is to find the right job for you.

Training for a Job

Most careers require specific skills. There are several ways to get these skills.

Occupational Training Centers

Most states have public **occupational** or **vocational training centers**. These centers offer programs in trades such as carpentry. Fig. 18-6. Those with a high school education who are interested in a trade can enter a training center after graduation.

Some centers offer adult education programs in the evening. These are for people who have daytime jobs and want to learn new skills.

Technical Schools

A technical school is usually privately owned. It offers training for specific careers. Programs range from one to two years in length.

Community Colleges

Community colleges are two-year schools that offer many programs. These schools often support the needs of local industries. They also offer general education courses. A person who graduates from a community college receives an associate's degree.

Four-Year Colleges

Many management positions in manufacturing and construction require a bachelor's degree. To get a bachelor's degree usually takes four years of study. General education classes, as well as a major field of study, are included.

▶▶▶ FOR DISCUSSION ◀◀◀

1. Do you know an adult who has gone back to school to get more education? What was that person's reason?

Extension
Activity

■ **Look in the yellow pages of your local phone book for colleges and training centers in your area. Choose one and try to learn more about it. If possible, visit your school or town library and ask to see a copy of the school's course catalog. Does it offer courses of interest to you? Report what you've learned to the class.**

What Career Is Best for You?

At this point in your life you may not have a clear idea of the kind of work you want to do. One field may interest you today. Tomorrow another one may sound more appealing. While you are still in school, you will be learning about yourself as well as the world around you. However, there are some things you can be thinking about that will help you later when you make a final decision.

Looking at Yourself

Most people choose a career based on their interests, abilities, aptitudes, and values. All of these are important in making a good career choice.

Your **interests** are the things you like to do or learn about. They may include your favorite school subjects, hobbies, sports, or even movies. Fig. 18-7. It's important to find a career field that interests you. Otherwise, you may spend many years of your life being bored or unhappy.

〰〰〰〰〰〰〰〰〰〰〰〰〰〰〰〰〰〰

Fig. 18-7. This young man is interested in cars. Would a career in car manufacturing appeal to him?

Your **aptitudes** are those skills you are capable of learning. For example, perhaps you have good eye-hand coordination. This means you may have an aptitude for welding. Welding is a skill you *could* learn.

Your **abilities** are those skills you have already learned. If you know how to use a power drill, then that is an ability.

Your **values** are your ideas and beliefs about things that you think are important. You may value such things as money, family relationships, security, expensive clothes, or independence. Your values can affect your happiness in a career. For example, suppose you valued wearing stylish clothes. Would you be happy in a job where you had to wear a work uniform all day?

Take a minute to think about your interests, aptitudes, abilities, and values. What do they tell you about the kind of career that would make you happy? Perhaps no job can satisfy all of them. However, keeping them in mind will help you make a wiser career choice.

Exploring Careers

Choosing a career is a form of problem solving. Getting the right information is important for making a good decision. The more you learn, the more you have to work with.

Talking to Others

One of the easiest ways to find out about different careers is to talk to the people who already work in them. Ask them about jobs that you think you would like. Most people are glad to share their experiences with someone who shows an interest. Fig. 18-8.

Research

Labor unions and professional organizations often provide career information. These groups usually have educational divisions that try to interest young people in their fields.

Your public or school library is another good source of information. The library will have a copy of the *Dictionary of Occupational Titles*. This book is published by the federal government and describes about 20,000 different jobs.

Another library reference is the *Occupational Outlook Handbook*, also published by the federal government. It tells you many facts about a particular job. Some of these include the type of work done, working conditions, and salary. It also predicts how many workers will be needed in that job in the future. Such information can help you determine which jobs will be plentiful when you're ready to go to work. Fig. 18-9.

Part-Time and Volunteer Jobs

Part-time jobs can also help you learn about a career field. Even delivering newspapers can give you a taste of responsibility and help you determine your aptitudes. Do you like keeping payment records or talking to customers best? Decisions like this may begin to give you some career direction.

Fig. 18-8. Many adults enjoy helping young people learn about careers.

Fig. 18-9. Your library is a good source of information about jobs. If you can't find what you're looking for, ask the librarian.

Volunteering is another way to get job experience. When you **volunteer** to do a job, you do not get paid for it. Many hospitals and community organizations need volunteers to do many kinds of jobs. School organizations may also need jobs done. Fig. 18-10. These jobs can also help you decide what your interests, abilities, aptitudes, and values are.

▶▶▶ **FOR DISCUSSION** ◀◀◀

1. **Have you ever talked to an adult about how that person chose his or her career? What did the person tell you?**

Extension

Activity

■ **Look at your local newspaper's job listings. Find ten jobs that relate to manufacturing and ten that relate to construction. Cut out the ads and make a poster or display of them. List the interests, aptitudes, values, and abilities a person might have who would find each of the jobs interesting.**

Fig. 18-10. This student is helping build a float for the homecoming parade. What do you think the job can teach her about her aptitudes, interests, abilities, and values?

Chapter Highlights

● Employers look for certain personal skills in every employee. The most important of these is a positive attitude. Other skills include the ability to follow directions and dependability.

● Basic skills learned in such courses as English and mathematics are an important part of a general education.

● Places to train for a job include: occupational or vocational training centers, technical schools, community colleges, and four-year colleges.

● Most people choose a career based on their interests, abilities, aptitudes, and values.

● When choosing a career, it helps to talk to others, to research the choices, and to work part-time or as a volunteer.

Test Your Knowledge

1. What is probably the most important personal skill to have for job success?

2. Name three ways to change a negative attitude.

3. What is the minimum education required for most manufacturing and construction jobs?

4. Name two courses that might be part of a general education.

5. What kind of centers offers programs in trades?

6. What is the difference between an associate's degree and a bachelor's degree?

7. What four things about yourself should you consider when choosing a career?

8. What are the skills you are capable of learning called?

9. Name two library reference books used to find career information.

10. What kind of work gives you job experience but no pay?

Correlations

LANGUAGE ARTS

1. Prepare a résumé. Include your name, address, telephone number, Social Security number, schools you have attended, jobs you have had, and three references.

SOCIAL STUDIES

1. Do some research on another country from Table C of the "Building a Future" activity on page 281. Suppose you are a bicycle manufacturer. Would you build a factory there? Why or why not?

SCIENCE

1. Identify an occupation you find interesting. Find out how science applies to that occupation.

MATH

1. Suppose you get a job in construction that pays $12.65 an hour. How much will you earn in one 40-hour work week?

.You•Can.

— Make a Difference —

By Helping Others, Students Become PALS

Through PAL — the Peer Assistance and Leadership program in Texas — students are learning that they can develop their own leadership skills and, at the same time, really make a difference in the lives of other young people. A high school student who is a PAL might tutor a first-grader in reading, work individually with a fifth-grader who needs extra help, or talk to someone else about the importance of staying in school.

Launched in 1980 in the Austin Independent School District, PAL is now active in hundreds of school districts throughout Texas.

In the program, carefully selected high school students first receive special training so that they'll have the knowledge and skills to help others. Students are chosen who can both reflect and respond to the student body as a whole.

In their training, the students learn to trust, to listen, and to communicate. They learn how to help other students solve problems and make informed and responsible decisions. They then use those skills at their own high school and/or at feeder elementary or middle schools. More and more programs are being set up in middle schools,

with older students helping younger students.

After training, PAL students help in many ways. At some schools, they welcome new students and make them feel at home. They tutor those who need extra help and assist others with special needs. At other schools, they meet with younger students to talk about such things as how to survive in high school, drug and alcohol abuse, how to deal with authority, and setting and reaching goals.

One PAL, a football player, wore his uniform to class one day and talked with younger children about rules, cooperation, and sportsmanship.

Another, a basketball player, helped first-graders sharpen their math skills as they kept a running tally of his successful baskets.

The peer helpers themselves gain many skills by working with the program. They learn patience, self-confidence, trust, and insight into the needs of others. The program has also helped them understand children better, which could make them become better parents in the future.

PAL students speak of the fulfillment they get from assisting other people, of realizing that they really can make a difference in another person's life.

The program has also received praise from those who have graduated and started careers. An attorney from the early days of the PAL program in Austin says he continues to use the communication, confidence, and people skills that he gained in the program.

In a recent survey in Texas, students were asked, "If you had a drug or alcohol problem and needed help, whom would you go to?" Seventy-three percent answered, "My friends." This is why programs like PAL work. Young people can be aided by those they trust — other young people.

Appendix

Appendix A. Measuring with the Metric System

Products made by American production systems must compete in world markets. Most other countries use the metric system of measurement. More and more American products are being made to metric sizes.

There are many measures used in the metric system. The important ones are explained here. (See also Table 1.) A handy conversion chart can be found in Table 2.

Metric Units

Length. The base unit for measuring length is the **meter (m)**, which is equal to about 40 inches. Another important unit is the **millimeter (mm)**, which is about as long as a dime is thick. All metric drawings for production are made to millimeter dimensions.

Area. The unit for area is the **square meter (m²)**, which is a little larger than the square yard. The **square millimeter (mm²)** is used for small area measurements.

Volume. The **cubic meter (m³)** is used to measure volume. It is slightly larger than the cubic yard. For measuring liquid volume, the **liter (L)** is used. It is a little larger than the quart. The **milliliter (mL)** is used for smaller volumes.

Mass or Weight. The metric base unit for mass is the **kilogram (kg)**, which weighs a little more than two pounds. The **gram (g)** is used for smaller weights.

Temperature. Temperature is measured in **degrees Celsius (°C)**. Important temperatures to know are:

100°C	Water boils
37°C	Body temperature
20°C	Average room temperature
0°C	Water freezes

TABLE 1. Metric Equivalents

Linear Measures
1 kilometer = 0.6214 mile
1 meter = 39.37 inches = 3.2808 feet = 1.0936 yards
1 centimeter = 0.3937 inch
1 millimeter = 0.03937 inch

1 mile = 1.609 kilometers
1 yard = 0.9144 meter
1 foot = 0.3048 meter = 304.8 millimeters
1 inch = 2.54 centimeters = 25.4 millimeters

Area Measures
1 square kilometer = 0.3861 square mile = 247.1 acres
1 hectare = 2.471 acres = 107,639 square feet
1 are = 0.0247 acre = 1076.4 square feet
1 square meter = 10.764 square feet
1 square centimeter = 0.155 square inch
1 square millimeter = 0.00155 square inch

1 square mile = 2.5899 square kilometers
1 acre = 0.4047 hectare = 40.47 ares
1 square yard = 0.836 square meter
1 square foot = 0.0929 square meter
1 square inch = 6.452 square centimeters

Cubic Measures
1 cubic meter = 35.315 cubic feet = 1.308 cubic yards
1 cubic meter = 264.2 U.S. gallons
1 cubic centimeter = 0.061 cubic inch
1 liter = 0.0353 cubic foot = 61.023 cubic inches
1 liter = 0.2642 U.S. gallon = 1.0567 U.S. quarts

1 cubic yard = 0.7646 cubic meter
1 cubic foot = 0.02832 cubic meter = 28.317 liters
1 cubic inch = 16.38706 cubic centimeters
1 U.S. gallon = 3.785 liters
1 U.S. quart = 0.946 liter

Weight Measures
1 metric ton = 0.9842 ton (long) = 2204.6 pounds
1 kilogram = 2.2046 pounds = 35.274 ounces
1 gram = 0.03527 ounce
1 gram = 15.432 grains

1 long ton = 1.016 metric tons = 1016 kilograms
1 pound = 0.4536 kilogram = 453.6 grams
1 ounce = 28.35 grams
1 grain = 0.0648 gram
1 calorie (kilogram calorie) = 3.968 Btu

TABLE 2. Approximate Customary-Metric Conversions

When you know:		You can find:	If you multiply by:
Length	inches	millimeters	25.4
	feet	millimeters	304.8
	yards	meters	0.9
	miles	kilometers (km)	1.6
	millimeters	inches	0.04
	meters	yards	1.1
	kilometers	miles	0.6
Area	square inches	square centimeters (cm^2)	6.5
	square feet	square meters	0.09
	square yards	square meters	0.8
	square miles	square kilometers (km^2)	2.6
	acres	square hectometers (hectares)	0.4
	square centimeters	square inches	0.16
	square meters	square yards	1.2
	square kilometers	square miles	0.4
	hectares (ha)	acres	2.5
Mass	ounces	grams	28.4
	pounds	kilograms	0.45
	tons	metric tons (t)	0.9
	grams	ounces	0.04
	kilograms	pounds	2.2
	metric tons	tons	1.1
Liquid Volume	ounces	milliliters	29.6
	pints	liters	0.47
	quarts	liters	0.95
	gallons	liters	3.8
	milliliters	ounces	0.03
	liters	pints	2.1
	liters	quarts	1.06
	liters	gallons	0.26
Temperature	degrees Fahrenheit	degrees Celsius	0.6 (after subtracting 32)
	degrees Celsius	degrees Fahrenheit	1.8 (then add 32)
Power	horsepower	kilowatts (kW)	0.75
	kilowatts	horsepower	1.34
Pressure	pounds per square inch (psi)	kilopascals (kPa)	6.9
	kPa	psi	0.15
Velocity (Speed)	miles per hour (mph)	kilometers per hour (km/h)	1.6
	km/h	mph	0.6

Metric Prefixes

The metric system is based on units of 10. Fractions are shown in decimals. Certain prefixes can be used with many units to show how they have been multiplied or divided by tens.

kilo = 1000
centi = ¹⁄₁₀₀ (0.01)
milli = ¹⁄₁₀₀₀ (0.001)

For example, if you know that the distance from school to town is 1000 meters, you would say "one kilometer." Instead of saying "¹⁄₁₀₀₀th of a meter," you would say "one millimeter."

Reading a Metric Rule

You read a metric rule the same way you read a customary rule. You count the markings. Look at the metric rule shown in Fig. A. It is marked in millimeters. To read it, count the number of millimeter spaces for a given length.

For example, the distance from A to B is 15 mm. The distance from A to C is 25 mm. What is the distance from A to D? A to E? A to F? Practice measuring paper clips, nuts and bolts, and other common objects. Round off your measurements to the nearest millimeter.

Designing in Metric

When designing a product to metric measurements, use a dual-reading metric rule. Think about replacement sizes, not conversions. For example, if you are designing bookshelves, a shelf width of 9 inches is about right. But now you want a metric size. Look at the dual-reading rule. You will see that 9 inches converts to about 229 mm. However, that is an odd size. It can be rounded off to 230 mm.

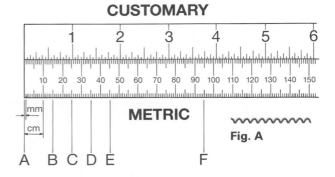

Fig. A

Because 230 mm is easier to read on your rule and more convenient to use, it's called a "rational" size. Try to select rational sizes ending in 5 or 0. Do the same for other measurements you want to use.

Changing Customary Sizes to Metric Sizes

Many product drawings are in customary inch sizes. You might want to change them to metric dimensions.

Study the customary inch-size bowl in Fig. B. Note the 5½-inch diameter. Now look at the dual reading rule. If you do not have a rule, you can multiply the inch dimension by 25.4 to convert it to millimeters. This is because there are 25.4 millimeters in one inch. It is easier to do this if you change the fraction to a decimal.

$$5.5 \times 25.4 = 139.7$$

The result can be rounded off to 140 millimeters. Change the diameter on the drawing to 140 mm. Convert the other dimensions.

If the size you need for a product is not one of the standard sizes, choose your materials in the closest standard size instead. For example, the drawing of the bowl calls for 22 gauge sheet metal. The nearest metric replacement size is 0.60 mm.

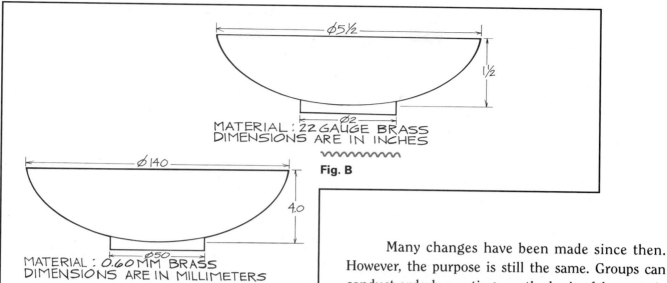

MATERIAL: 22 GAUGE BRASS
DIMENSIONS ARE IN INCHES

Fig. B

MATERIAL: 0.60 MM BRASS
DIMENSIONS ARE IN MILLIMETERS

Appendix B. How to Conduct a Meeting

At some time during this course, your class will form a student company. Your company will then manufacture or construct a product. In Chapter 2, "Forming a Company," you will learn how a company is started and run. One thing that often takes place within any company is meetings. Meetings allow workers to communicate with one another. At meetings, most decisions are usually made.

People have used meetings as a way to get things done for many centuries. Over time they found that they needed rules for the meetings. Without rules, some people never got to speak. Others did all the talking. To have a discussion in which all could participate, the group needed a leader who would guide the discussion. This leader had to know the rules for discussions. In England, Parliament began to use these rules when conducting business. The rules became known as parliamentary procedure.

Many changes have been made since then. However, the purpose is still the same. Groups can conduct orderly meetings on the basis of democratic principles. *All* members may contribute ideas and opinions, not just a select few.

The club meetings described here are fairly formal. Most businesses are more relaxed about meetings. However, the basic procedure is the same. There is one other important difference. In some businesses, final decisions must be approved by a supervisor.

Rules of Order

Parliamentary procedure is based on the following rules:

- During a debate, group members must be fair and polite.
- Any member of the group may debate under the rules established.
- The majority (side with the most votes) decides the issues.
- The minority (side with the least votes) is free to express its opinion.
- Minority members must go along with the decision made by the majority.

Terms to Use

To participate in groups using parliamentary procedure, members must know certain terms. Only then can the meeting move ahead smoothly.

- **A quorum:** the number of members who must be present for the group to do business and make decisions.
- **Minutes:** a written record of the business covered at a meeting.
- **Majority:** at least one more than half of the members present.
- **Motion:** a suggestion by a member that certain action be taken by the group.
- **Second the motion:** a second member approves the suggestion.
- **Amendment:** a change in a document.

Conducting a Meeting

No official business can be done at any group meeting unless a quorum is present. Usually a quorum is a majority of members. It can also be a number stated in the rules of the group. When a quorum is present, the meeting usually proceeds like this:

The group leader calls the meeting to order. Then he or she presents business in the order stated in the group's rules. If an order of business has not been agreed upon, the group should use the order described below. The club meeting that follows would be typical for many groups.

The president would make these statements:

"The meeting will come to order. Marie Smith, the secretary, will read the minutes of our last meeting."

"Are there any corrections to the minutes?" (If no corrections are made, the meeting proceeds.)

"The minutes are approved as read." (If corrections are needed, the president asks the secretary to correct the report.) "Are there any corrections?" (pause) "Then, the minutes stand approved as corrected."

"The next order of business is the reports of other officers." (Examples would be a secretary's report or a treasurer's report.)

"The next order of business is the reports of committees."

"Is there any unfinished business to act upon today?"

"Is there any new business to be considered today?" (New business includes communications, presenting bills for payment, future plans, setting dates for other activities, etc.)

"Is there future business or any announcements?" (After announcements are made, the president may end the meeting.)

Motions

In a motion, a member suggests that the group take a certain action.

Any member may make a motion to introduce an item of business. A motion must be seconded. After it is seconded, the motion may be discussed. Finally, a vote will be taken. The majority always rules. An example follows:

John stands and addresses the president, "Mr. President."

The president recognizes John by calling his name.

John says, "I move (make a motion) that the club buy a new ceremonial emblem before the state contest in April."

Carol, another club member, agrees, "I second the motion."

Then the president announces, "It has been moved and seconded that the club buy a new ceremonial emblem before the state contest in April."

At this point, the other members discuss the motion.

Finally, the president asks, "Is there further discussion?" After a pause, he prepares for the vote. "All in favor of the motion say 'Aye.'" He counts the favorable votes. "Those opposed say 'No.'" He counts. "The ayes have it. The motion has passed that we buy a new ceremonial emblem before the state contest."

Electing Officers

The democratic process requires that persons in leadership positions be nominated and elected by the group. Nominations are made by the members, by a committee, or by ballot.

In many organizations, new officers are nominated by a committee appointed by the president or elected by the members.

On the date set for election of officers, the president asks for a report from the nominating committee. Then, the persons nominated are named and a vote taken.

Glossary

This glossary defines terms used in production. Some of the terms have additional meanings, but only the definitions that relate to manufacturing or construction are given here.

A

ability (uh-BILL-uh-tee) A skill that is learned.

abrasion The process of rubbing chips of a material away with a harder material called an abrasive.

acceptance quality control Examination of materials, parts, or finished products at certain points to make sure specifications have been met.

acid rain Precipitation that is created when sulfur and nitrogen oxides combine with water in the atmosphere to form sulfuric acid.

adhesion (uhd-HE-zhun) The use of glue or other adhesives to hold parts together.

affirmative action program (uh-FUR-muh-tivv) Program designed to bring more minority members into a company's workforce.

airspace construction Building structures above and below highways.

alloy (AL-oy) A metal that is a mixture of two or more metals or of a metal and a nonmetal.

alterations (ALL-tuh-RAY-shuns) Changes made to the structural form of a building or other structure.

applied research Also called development, this describes what occurs when scientists and engineers take the information discovered in basic research and try to find ways to apply it.

apprenticeship (uh-PREN-tihs-ship) Training program that combines formal classroom training with on-the-job training by a skilled worker.

aptitude (APP-tuh-TOOD) The capacity to learn.

aquaculture (AHK-wuh-KULL-chur) Controlled production of plants and animals that live in or upon the water for human use.

arch bridge A curved bridge in which the arch carries the weight of the load. The arch is supported at each end by an abutment.

artificial intelligence Programs that enable computers to learn from previous experience and apply that knowledge to new situations.

assembly drawing Drawing showing how all the parts are put together to make a finished product.

assembly line In factories, an arrangement by which each worker does a single operation in assembling the product as it is passed along, often on a slow-moving belt or track.

associate degree (uh-SOH-she-it) *or* (uh-SOH-see-it) Degree received upon completion of a two-year college program.

attitude How one views life and works with other people.

B

bachelor's degree Degree awarded upon completion of work at a four-year college.

bar chart Chart comparing various amounts by the length of thick, solid lines or bars.

bar codes A series of black lines representing numbers that computer scanners can read. They are used to help maintain inventories.

basic research Research done to learn new information.

batter boards Boards held horizontally by stakes driven into the ground to mark the boundaries of a building.

beam bridge Bridge in which steel or concrete girders (long, supporting beams) support a concrete roadway.

bearing-wall superstructure Structure in which heavy outer walls support the weight of the structure.

bid A price quote, usually in writing, that states how much a contractor will charge for a project.

bioconversion (BY-oh-kun-VERR-zhun) The process of using plant and animal wastes to produce fuels.

biomass energy Energy that comes from plant and animal wastes. A wood fire is one example.

biotechnical Having to do with the use of knowledge about biology and engineering to solve human problems.

biotechnology The use of knowledge about biology and engineering to solve human problems.

blow molding Process in which heated and softened plastic is placed in a mold. Then compressed air is blown into the plastic to force it against the sides of the mold.

bonds Certificates sold by corporations as a way of raising money; they promise to return the money, along with interest, to the buyer by a certain date.

brittle Describes a material that cannot be deformed (changed in shape) without breaking.

building codes Local and state laws that specify the methods and materials that can be used for each aspect of construction.

building permit Document issued by the local government that indicates a structure is approved for construction.

C

cable-stayed bridge Bridge supported by cables that are connected directly to the roadway.

CAD *See* **computer-aided design**

CAM *See* **computer-aided manufacturing**

Canadarm A fifty-foot robotic arm used on the space shuttle to launch and repair satellites.

cantilever bridge (KAN-tull-EE-vur) Bridge consisting of two beams, or cantilevers, that extend from each end and are joined in the middle by a connecting section called a suspended span.

cantilever A projecting beam supported only at one end and used in bridge design.

capital Money used to pay for business or personal needs.

cash flow The pattern of expenses and income of a company. If a company brings in more than it must pay out, it has a positive cash flow. If it must pay out more than it brings in, it has a negative cash flow.

casting Process in which materials are softened or melted and then forced or poured into a mold. As it solidifies, the material takes the shape of the mold.

certificate of occupancy (AHK-yuh-pun-see) A certificate issued by the building inspection department indicating the building has passed the building inspector's check and is safe for people to use.

CIM *See* **computer-integrated manufacturing**

circuit breaker Switch in the service panel that opens automatically to break the flow of electricity when there is an electrical overload.

civil construction projects Projects built for public use, such as highways, bridges, and tunnels.

clear-cutting Method of harvesting timber in which every tree in the forest is cut at the same time.

climate control systems Heating, air conditioning, and ventilation systems in a building.

CNC *See* **computer numerical control**

cofferdam (COUGH-ur-DAM) A watertight, temporary dam built to shift the flow of water around the enclosed area so that construction can be done on a permanent dam.

cogeneration (KOH-jenn-uh-RAY-shun) When a single energy source is used for two purposes.

cohesion (Koh-HE-zhun) The joining of two materials by melting them together.

collective bargaining The negotiation of working conditions, salary, and benefits between an employer and a group of workers, who are usually represented by a labor union.

combining The process of joining or adding materials together.

commercial buildings Buildings that house businesses, such as supermarkets and shopping malls; may also include churches and public buildings.

composite (Kum-PAHZ-it) New material made by combining two or more materials.

compression (Kum-PRESH-un) Forming process in which the material is squeezed together to form the desired shape.

compressive strength The ability of a material to resist being squeezed or pressed.

computer-aided design (CAD) Using a computer to assist in the creation or changing of a design.

computer-aided manufacturing (CAM) Using computers to instruct machines on how to manufacture parts.

computer-controlled tools and machines Machines whose operation is controlled by computers.

computer-integrated manufacturing (CIM) (IN-tuh-GRATE-id) All areas of a company (finance, manufacturing, marketing, etc.) are linked through computers so each department can keep informed of other departments' progress or problems and adjust their procedures accordingly.

computer-numerical control (CNC) (NEW-mair-ih-cull) An operator programs a computer to automatically control a machine's operations.

conditioning Manufacturing process in which the internal structure of a material is changed in order to improve its properties.

constructed features Features that add to the appearance or function of a building but are not actually part of it. Examples include patios, driveways, walks, fences, and decks.

construction Using manufactured items to build something on site.

construction boss The person in charge of everything that takes place on a construction site.

consumers Buyers of goods and services.

conveyor belt A continuous belt driven by an electric motor and used to transport items from one place to another.

corporation (KORR-puh-RAY-shun) A company that is owned by many people (stockholders) but is allowed by law to act as a single person.

corrosion (Kuh-ROH-zhun) The gradual eating into or wearing away of a material caused by the action of acid or rust.

cranes Heavy equipment used to lift and swing loads from side to side by means of a long cable that is wound up on a pulley (hoist).

critical defects Defects in a product that place the user of the product in danger.

critical path method A diagram made of circles and lines showing the sequence of critical parts of a manufacturing project.

custom manufacturing Process in which a product is made in limited quantities, sometimes to a customer's specifications.

customize Adding unique features to a product to make it different from others of its kind.

D

dead load The basic weight of a structure.

debt Money that has been borrowed from and is owed to an individual, business, or lending institution.

Delphi Study (DELL-feye) Survey in which a panel of experts forms conclusions on questions about the future.

detail drawing Drawing of a particular part of a floor plan or elevation drawn to a larger scale and giving more information than the original plan.

Dictionary of Occupational Titles U.S. government publication that describes about 20,000 different types of jobs relating to many different careers.

die casting Molten metal is forced into a mold (die) under pressure; as it solidifies, it takes on the shape of the die.

dies Molds made of steel that are used to form materials.

direct mailings Advertisements mailed directly to consumers' homes.

distribution (DISS-truh-BYOO-shun) The moving of products from the factory to the customer.

dividends (DIVV-uh-dends) Amount of money from a company's profits paid to stockholders on each share of stock.

drawing Forming process in which heated metal is pulled or stretched through a die to take on the desired size and shape.

drywall A common material used to cover ceilings and walls; usually made of a plaster-like substance covered with paper.

ductility (duck-TILL-uh-tee) The ability of a material to be changed in shape without breaking or fracturing.

E

elasticity (ih-LASS-TIHS-uh-tee) *or* (EE-lass-TIHS-uh-tee) The ability of a material to readily return to its original shape after a force has been applied and then removed.

electric power Power that comes from electricity. A motor changes the electricity into a force that can do work.

elevations (EL-uh-VAY-shuns) Drawings that show the outside of a structure, as viewed from the ground level. Separate elevations are done for each side of a structure.

energy The ability to do work.

entrepreneur (AHN-truh-pruh-NUR) A person who starts his or her own business.

Environmental Impact Statement (inn-VY-run-MEN-tull) A report detailing how a proposed construction project will affect the area, including inhabitants, surrounding it.

Environmental Protection Agency (EPA) Federal agency that identifies danger to the environment and enforces environmental laws.

EPA *See* **Environmental Protection Agency**

estimate A calculation of how much a construction job is going to cost and how long it will take to complete.

excavation The process of digging a hole for the foundation of a structure.

excavators Heavy equipment used to dig and move dirt when preparing a site for construction.

expenses The total amount of money required to make a product.

extrusion (ik-STROO-zhun) Forming process in which heated metal or plastic is forced through a die (or opening) of the desired shape.

F

fatigue strength (fuh-TEEG) The ability of a material to withstand repeated loads.

feedback Information about the output (results) of a system that is sent back to the system; can be used to improve the performance of the system.

field engineer The person who oversees large construction projects, such as highways or skyscrapers.

fixture (1) A device used to hold a part while it is being machined or assembled. (2) A permanent device inside a building, such as a ceiling light or bathtub.

flat paints Paints that leave a dull surface when dry.

flexible manufacturing systems (FMS) Using computers to produce small quantities of a variety of similar items using the same machines.

floor plan Diagram showing the arrangement of a room or rooms as viewed from above; shows walls, doors, windows, stairs, built-in cabinets, and other features.

flow chart A diagram that identifies all the operations needed to make a product.

FMS *See* **Flexible manufacturing systems**

footing The part of a structure below a foundation wall that distributes the structure's weight.

forming The process by which the shape of a material is changed without adding or taking anything away.

fossil fuels Fuels that come from plants and animals that lived millions of years ago. Coal, petroleum, and natural gas are fossil fuels.

foundation (fown-DAY-shun) The part of the building that is below the first floor; it is designed to support the superstructure.

frame superstructure Structure in which many connected parts support the weight of the building.

free enterprise system (EN-terr-PRIZE) System in which most goods and services are provided by privately-owned, for-profit companies.

frost line The depth to which the soil freezes in winter. Footings must be located below the frost line.

futurists (FEW-churr-ists) People who study the future and make predictions about it based on analysis of past events and on study of current and planned research.

G

gages Measuring instruments used to check the size of parts.

general contractor A construction company that agrees to manage an entire project and be responsible for all the work done from start to finish.

genetic engineering (juh-NET-ick) The blend of science and technology used to study and change DNA, which contains the genetic codes that determine the characteristics of all living things.

geothermal energy (GEE-uh-THUR-mull) Energy that comes from within the earth's interior.

geyser (GUY-zur) A column of hot water and steam that shoots up from the earth.

gloss paints Paints that leave a very shiny surface when dry.

grading Giving the land the shape called for in a landscape plan.

H

hand tools Tools that are easily held in the hand and are powered by the person using them. Examples include hammers, screwdrivers, and handsaws.

hardness The ability of a material to resist scratching and denting.

hardwoods Woods that come from trees that lose their leaves in the fall.

heavy construction Building methods and materials used for large structures, such as tunnels, factories, and high-rise buildings.

human resource management Department of a company that concerns itself with recruiting, hiring, training, advancing, and separating employees; also called the personnel department.

hydraulic power Power that comes from the pressure of a liquid, usually oil. The liquid is forced through hoses to the area where the force is needed.

hydroponics (HI-druh-PAHN-icks) The growing of plants without soil.

I

impact strength The ability of a material to withstand sudden impact, such as a hammer blow; also called toughness.

impacts The effects something has on something else.

industrial buildings Buildings used by production industries.

injection molding (in-JECK-shun) The process of forcing heated, softened plastic under pressure into a mold of two halves clamped together. When the plastic cools and solidifies, the halves of the mold separate to release the finished plastic product.

input Anything that is put into a system. All technological systems require input from the following types of resources: people, information, materials, tools and machines, energy, capital, and time.

inspector A worker who checks for flaws in a product or structure.

insulation Material used to keep heat from penetrating a building in summer and cold from penetrating in winter; also keeps heated or cooled air from escaping outdoors.

interchangeable parts Parts that are made exactly like one another so that any one of them will fit the product.

interest (IN-trist) *or* (IN-tur-ist) The fee charged by a lender for the use of capital (money).

interests The things a person likes to do or think about. Examples include hobbies, favorite school subjects, and sports.

intermittent manufacturing (IN-ter-MITT-nt) Manufacturing operation in which mass production methods are used to produce a certain number of units over a limited period of time. Production may run in cycles with one product being made for a few weeks and then the production line changed over for a different product.

inventory (IN-vun-TORE-ee) The items a company has in stock.

inventory control Management technique that keeps track of stock on hand.

J

jig A device used to hold a part and guide a cutting tool.

JIT *See* **just-in-time**

jointing Using a cutting tool (jointer or plane) to make the edge of a piece of wood square and smooth.

just-in-time (JIT) A method of inventory control in which no items are warehoused. Materials are delivered by suppliers just in time to be used in making the product.

K

kinetic energy (kih-NET-ik) The energy of motion. The energy of falling water used to produce electricity is one example.

L

landscaping Grading the earth around a structure to give it the desired contour, constructing driveways and walks, and planting trees, shrubs, grass, and flowers.

laser (LAY-zur) A narrow, intense beam of light that concentrates a lot of power in a very small space.

layout tools Tools used to create straight lines, angles, and circles when drawing plans. Examples include a compass and a framing square.

lead time (leed) Amount of time required between ordering of materials and their delivery.

lean production Approach to manufacturing in which workers are organized into teams that determine the best way to perform their assigned tasks.

lease A long-term rental in which a monthly fee is paid for the use of equipment for a specified period of time.

light construction Building methods and materials used to make smaller structures, such as single-family homes and garages.

live load The weight applied to a structure, including people, furniture, snow, and so on. Live loads often change.

load-bearing walls The walls that help support a structure. They may be exterior or interior walls.

loan Borrowed money that will be repaid with interest.

locks Special chambers built into canals that allow a ship to be raised or lowered on its way through. Locks are needed when land or water levels change along the canal route.

M

mainframe computer Room-sized computer to which many work stations are attached.

maintenance (MAIN-tun-uns) (1) The scheduled servicing and emergency repairs of tools and machines. (2) Taking care of a structure and its surrounding property.

major defects Defects that could reduce a product's life span or cause its operation to fail.

management The act or skill of controlling or directing activities. Also, those who control or direct activities in a company.

manufacturing Creating products in a factory or some other processing facility.

marketing The methods a company uses to sell its products.

market research Method of determining exactly what customers need and want. Two market research techniques are telephone surveys and mailed questionnaires.

markup The difference between the original purchase price and the selling price when a product changes hands from manufacturer to wholesaler to retailer to consumer.

mass production The process of making large quantities of a product on assembly lines using interchangeable parts.

mass superstructure Structure in which large quantities of materials are used to construct a solid or nearly solid structure. Examples include stone monuments, pyramids, and dams.

measuring tools Tools used to determine the size of an object. Examples include a folding rule and calipers.

mechanical fastening Using metal or plastic devices to hold parts of a product together. Examples of fasteners include nails and screws.

mechanical properties The way a material reacts to loads and forces. Mechanical properties include elasticity, plasticity, ductility, brittleness, tensile strength, compressive strength, shear strength, fatigue strength, hardness, and impact strength.

microgravity (MY-crow-GRAVV-uh-tee) The very low gravity that exists on board the Space Shuttle.

milling Cutting by moving the workpiece past a tool having sharpened teeth equally spaced around a cylinder or along a flat surface; done to smooth a surface or cut grooves or slots.

minor defects Small defects that do not affect the operation or usability of a product.

modular construction (MAHJ-ur-lur) Type of construction in which sections of a building are made at a factory and are then shipped to the building site, where they are assembled.

motion study An evaluation of a worker's movements to determine the best sequence (order) for performing a task.

movable bridge Bridge with a section that can be tilted or lifted upward to allow large ships to pass underneath.

N

NASA The National Aeronautics and Space Administration, which plans and operates the American space program.

natural materials Materials that are found in nature. Examples include wood, many metals, coal, oil, and natural gas.

navigation canals Canals used by ships for transportation.

needs Those things that are required to keep us alive and well.

needs assessment Gathering of information to evaluate the need for a new structure.

negative attitude A way of looking at something so that its unhappy aspects are stressed.

nuclear energy Energy produced by splitting the atoms of elements such as uranium.

O

Occupational Outlook Handbook U.S. Government publication that forecasts future demand for jobs and other information about several hundred careers that are grouped into large general categories, or clusters. Other information includes duties and responsibilities, training required, wages, working conditions, and opportunities for advancement.

Occupational Safety and Health Administration (OSHA) The division of the U.S. Department of Labor that sets standards for health and safety in the workplace and sees that these standards are met.

occupational training center Public training facility that offers programs in trades, such as carpentry.

on-the-job training Employees are trained by an experienced worker or supervisor as they perform their regular work activities.

optical qualities The way a material reacts to light.

OSHA *See* **Occupational Safety and Health Administration**

output The result of a system; the amount of goods and services produced by a production system.

oxidation (AHK-suh-DAY-shun) Effect of the oxygen in air mixing with and acting upon substances. In metals containing iron, oxidation causes rust.

P

packaging The container into which a finished product is placed.

panelized construction (PAN-ull-ized) Construction in which wall panels are built at a factory to the builder's specifications and are then shipped to the construction site where they are assembled to produce the framed and sheathed walls.

partitions (parr-TISH-uns) Nonload-bearing walls that divide a building into rooms.

patent (PAT-unt) The exclusive right, granted by the U.S. government, to produce something.

pension Money paid regularly to a person who is retired or unable to work because of a disability.

photovoltaic cells (FOAT-oh-vohl-TAY-ik) Devices that convert energy from the sun's light directly into electricity. Also called solar cells, these are often used in calculators and wristwatches.

planing (PLAY-ning) Cutting process done to flatten and smooth the face of a piece of wood.

plantings The trees, shrubs, grass, flowers, and other plants that surround a building.

plasticity (plas-TISS-uh-tee) The ability of a material to keep its new shape after it has been deformed (changed in shape).

pneumatic power (new-MAT-ik) Power that comes from air pressure and used to move the parts of a tool or machine.

pollution (puh-LOO-shun) The contamination of something by unwanted substances.

positive attitude A way of looking at something so that its happy aspects are stressed.

potential energy (puh-TEN-shull) Energy that is stored or at rest but that can become usable under the right conditions. One example is the energy stored in a battery.

power tools and machines Tools and machines that are powered by electricity, hydraulics (liquid pressure), or pneumatics (air pressure).

preventive maintenance (MAIN-tun-uns) Maintenance or servicing done on a regular basis to anticipate and correct problems before they occur and to increase lifetime.

primary processes Processes that convert raw materials into industrial materials.

primer A coat of paint that seals the wood or other surface to which it is applied. A finish coat is applied later.

principal The amount of money owed for a loan or a bond, not counting the interest.

problem-solving process A six-step process that can be followed to develop solutions to problems and school tasks. The six steps are: state the problem clearly; collect information; develop alternative solutions; select the best solution; implement the solution; evaluate the solution.

process The action part of a system; operation that changes the size, shape, or appearance of a material.

process quality control Process by which workers, methods, and machines are inspected throughout the manufacturing process to make sure the product is made correctly.

production engineering The activities involved in designing and building manufactured items.

production quota The number of items that must be produced each month.

production schedules Show when certain stages of an operation should begin and specify when raw or industrial materials are needed in the production process.

production system System in which each part has a specific role and all the parts work together to manufacture or construct products. These products include such things as clothes, appliances, books, and buildings.

production workers Manufacturing workers who change the form of materials to help create the finished product. They may be involved in such processing activities as cutting, welding, painting, or assembling.

productivity (PRO-duck-TIVV-uh-tee) The measure of the amount of goods and services produced (the output) and the amount of resources (the input) that produced them.

profit The amount of money a business makes after all expenses have been paid.

prototype (PROHT-uh-TYPE) A full-size working model of an actual product. Usually, a prototype is the first of its kind and serves as a model for others.

punch list An inspector's list of corrections that must be made to a structure.

purchasing agent Person in charge of selecting materials that meet specifications and making sure they are delivered in the right amount and in time to meet production needs.

Q

quality assurance (uh-SURE-uns) (1) Monitoring workers and machines as well as checking parts and finished products to make sure the product meets all design specifications and customer expectations. (2) Monitoring to make sure methods and materials used during construction meet all building codes and design specifications and the work is done carefully and skillfully.

quality circles Meetings in which workers discuss production problems and offer suggestions for improvement.

R

R-value Measurement indicating a material's ability to insulate.

random sampling Technique in which only a certain percentage of the completed parts in a production run is tested for defects.

recycling Using old, discarded products or scrap and waste materials to make new products.

remodeling Making changes in or to a building to make it more suitable to its current use or more attractive.

renewable resources Natural resources that can be replaced. For example, when trees are cut down for lumber, new trees can be planted in their place.

renovation (RENN-uh-VAY-shun) Repairing and restoring a building or other structure to return it to the way it originally looked.

repairs Process of fixing damaged areas in a structure.

research and development department Manufacturing department responsible for coming up with new or improved products and methods.

residential buildings (REZZ-uh-DEN-shull) Buildings in which people live.

resource A supply of something to take care of a need; anything that provides support or supplies for the system.

restoration (RESS-tuh-RAY-shun) Process of returning a building or other structure to the way it originally looked.

résumé (REZZ-oo-MAY) *or* (RAY-zoo-MAY) A brief report that includes information about a job applicant's education, experience, and desired employment.

retailers Those who sell products directly to consumers.

retort pouches (REE-tort) Food packages made of aluminum foil sandwiched between layers of plastic to protect the food from light, moisture, and air. Food in retort pouches can be stored without refrigeration for years.

rough work Placement of utility system pipes or wires inside the walls of a structure before the walls are covered.

routing (ROWT-ing) Process in which a revolving cutter removes material to shape or cut a piece. A router is often used to make decorative edges and to cut grooves or lettering into boards.

S

salaried employees Workers who are paid weekly or biweekly based on a yearly salary, rather than an hourly rate.

sand casting Molding method in which a pattern is packed into damp sand and then removed. Molten metal is poured into the impression left in the sand and then allowed to cool and solidify before removing.

SBA *See* **Small Business Administration**

scenarios (sih-NAIR-ee-OHS) Brief stories that outline situations or a series of events.

schedule A report showing when different stages of an operation should begin.

seconds Products that have minor defects that may affect the appearance but not their usability or operation.

section drawing Drawing showing a slice of the inside of a structure or object.

semigloss paints Paints that leave a slightly shiny surface when dry.

sensory properties Those properties we react to with our five senses — sight, hearing, touch, taste, and smell.

separating (1) A process in which materials are cut to a desired size and shape. (2) Ending employment; may be temporary, such as in some layoffs, or permanent.

service contract A written agreement with a firm that sells equipment, such as a heating and cooling system; the firm agrees to maintain the equipment.

service panel A box that contains circuit breakers or fuses for each individual, or branch, electrical circuit in a building as well as a main circuit breaker that controls all the circuits collectively.

servicing Supplying parts and labor to correct problems with a product.

shares A portion of ownership in a company.

shear strength Measurement of how strong a material is when it is being pushed in opposite directions at the same time.

shearing Cutting materials to the desired size or shape using blades or dies; may not involve loss of material.

sheathing (SHE-thing) Layer of material, such as plywood or fiberboard, that is placed between the framing and the finished exterior of a structure.

siding The outer covering on a framed superstructure. Siding may be bricks, wood, vinyl, or aluminum.

simple machines Early muscle-powered machines that serve as the basis for all the machines we use today. They include the wedge, the wheel and axle, the pulley, the screw, the lever, and the inclined plane.

simulation (SIMM-you-LAY-shun) Using one system to imitate another, such as using a computer to simulate a natural situation.

simulator (SIMM-you-LATE-er) A training device that artificially duplicates the conditions likely to be encountered in some operation, such as flying an airplane.

site The land on which a project will be constructed.

site plan A diagram showing the location of the building on the lot, boundaries, roads, the contour of the earth, and landscaping.

Small Business Administration (SBA) An agency of the federal government that gives financial aid to small businesses.

smart buildings Buildings in which computer systems control utilities, security systems, and appliances.

smelting The process of melting metal ore in order to get the pure metal away from any impurities or waste matter.

softwoods Woods that come from cone-bearing trees; many of these trees keep their leaves (needles) all year long.

solar cells Devices that convert energy from the sun's light directly into electricity.

specialty stock Material made to the buyer's specifications.

specifications (SPESS-uh-fih-KAY-shuns) (1) Detailed descriptions that give all the information needed to make a product. (2) Written details about what materials are to be used for a structure and the standards and government regulations that must be followed.

spinoffs Products or processes that are created as a result of the space program's technologies but are for use on earth.

standardized movements Guidelines indicating the best sequence of motions for workers performing particular operations.

standard sizes The most popular sizes of materials, such as fasteners and lumber, that are used to make products or structures.

standard stock Any material that is produced in common sizes and is readily available from various suppliers.

stock A share of ownership in a company.

storage The holding of products until they are used or purchased by customers.

stress Forces created inside a material by other forces acting on it from the outside.

strike A refusal by union members to work until they reach an agreement with the company.

subassemblies Smaller units put together to make a final product.

subcontractor A construction company that specializes in one type of work, such as plumbing or painting.

subsystem One of several smaller systems that combine to produce larger systems.

superstructure The part of the building above the foundation, beginning with the first floor.

supply and demand Describes the relationship between what customers want (demand) and the amount of a product a manufacturer can provide (supply).

surveying Measuring done to determine the exact boundaries of a piece of property.

suspension bridge (suh-SPENN-shun) Two tall towers support main cables, which run the entire length of the bridge and are secured by heavy concrete anchorages at each end. Suspender cables dropped from the main cables are attached to the roadway.

synthetic materials (sin-THET-ick) Materials that are created by people; they do not occur naturally. Examples include nylon and plastics.

system An organized way of doing something.

system drawing Drawing showing plumbing, electrical, or heating and ventilating systems.

T

technical school School that offers training for specific careers. Schools usually have a one-year or two-year program and charge tuition.

technological systems (TEK-nuh-LODGE-ih-kull) Systems made by people to satisfy human wants and needs. Examples are transportation systems and communication systems.

technology Applying the principles of science to do something; using this knowledge to make products that meet people's needs and wants.

tensile strength (TEN-sul) The ability of a material to resist being stretched or pulled apart.

test marketing Trying out a new product on a small number of consumers to learn their opinion.

thermal Having to do with heat.

thermoplastics Plastics that can be softened with heat and molded to a desired shape. Since they can later be reheated and remolded, scrap materials can be recycled.

thermoset plastics Plastics that are set into permanent shape by heat and cannot be reshaped.

time A measured period in which something happens.

time and motion study A measurement of the time it takes to complete each operation in a production process.

tissue culture Method of growing whole plants from tiny pieces of parent plants that are placed in a tube containing a growing medium.

tolerance (TALL-ur-uns) The amount of error in size that can be allowed for a part.

tooling up The process of setting up, servicing, and building the necessary tools, machines, and other equipment needed to make a product.

torsion strength (TORE-shun) The ability of a material to resist twisting.

toughness The ability of a material to withstand sudden impact, such as a hammer blow; also called impact strength.

trend analysis Involves reviewing newspapers, magazines, and research reports to learn which topics are frequently written about in order to identify important trends.

truss Preassembled triangular frame of wood or steel designed to frame and support the roof of a structure.

truss bridge Bridge supported by steel or wooden trusses, which are beams put together to form triangular shapes.

turbine (TERR-bin) *or* (TERR-byn) A wheel with blades or vanes attached to it at even intervals; the pressure of water or steam hitting the blades turns an attached shaft, which in turn drives a generator to produce electricity.

turning The process of revolving a workpiece against a sharp cutting tool to get the desired shape.

U

universal systems model A model system that can be applied to all technological systems and many natural systems to help the user understand new and unfamiliar systems.

utility systems (you-TILL-uh-tee) The electrical, plumbing, and climate control (heating and cooling) systems of a building.

V

values A person's ideas and beliefs about things the person thinks are important.

vocational training center *See* **occupational training center**

volunteer To offer to do a job without being paid for it.

W

wants Things that enable us to go beyond our basic needs to make our lives easier or more pleasant.

warehouse Large building in which products are stored until customers order them.

warranty (WAR-un-tee) Manufacturer's written promise to repair or replace, during a specified period of time, a product that does not work or perform correctly.

wholesaler Company that purchases products from various manufacturers and then resells them to retailers.

work Moving a force through a distance.

working drawings Drawings containing the information needed to build a product or structure.

X, Y, Z

zone of acceptance The area within a part's stated tolerance.

zoning laws Regulations telling what kinds of structures can be built in certain areas of a community; may also specify minimum property sizes, etc.

Photo Credits

AP/Wide World Photos, Inc., 50, 175, 201, 204, 209, 211, 282, 284, 290, 291

Adept Technology, Inc., 5

Adventure Photo/Eric Sanford, 17

Courtesy of the Arizona Historical Society/Tucson, 217

Arkansas Dept. of Parks & Tourism, 213

Arnold & Brown, 28, 41, 47, 77, 96, 98, 157, 158, 164, 167, 168, 182, 196, 199, 242, 265, 287

James L. Ballard, 309

Banana Kelly, 276-277

Roger B. Bean, 27, 30, 41, 54, 90, 181, 252, 263, 317

Keith Berry, 52, 198, 308, 313

Bethlehem Steel, 49, 160

The Bettmann Archive, 140

Black & Decker, 118, 119, 120

Black Diamond Equipment, Ltd., 17

David Edward Byrd, 15

CS&A, 44, 81, 134

Carr Lane, 153

James Carlson, 269

Caterpillar, Inc., 6, 210

Center for Disease Control, Atlanta, GA, 53

Courtesy of the Chicago Convention and Tourism Bureau, 229

Chrysler Corporation, 56, 142, 301

Cincinnati Milacron, 122, 144, 152

Circle Design, 279, 280, 281, 282, 283, 284, 285
 Carol Spengel, 12, 70, 132, 200, 278

Courtesy Coachmen Industries, Inc./Joe Hilliard, 139

Coherent General, Inc., 166

Comstock, 278

Conagra, Inc./Rick Harrig, 317

Consumers Union, 198, 199

Crawford, Murphy & Tilly, Inc., 217

The Daily Eastern News/James Carlson, 318

Howard Davis, 114, 115, 116, 117, 119, 216, 228, 234, 268

Deere & Company, Moline, IL, 92, 141

Delta International, 118 , 119

Denton Bicycles/Ann Garvin, 141

Diebold, Inc., 173

Walt Disney World Dolphin Hotel, 224

DoAll Company, 149

Dresser Industries, Inc., 110, 194

E-Systems, Inc., 9

Emulsion Polymers Institute, Lehigh University/NASA, 289

FAA Technical Center, 294

Randy Feucht Construction, Inc./Ann Garvin, 240

Fina Oil & Chemical Co., 193

Flow Systems, 165

Ford Motor Company, 19, 300

Fox & Jacobs, 268

David R. Frazier, Photolibrary, Inc., 20, 28, 84, 87, 103, 104, 153, 190, 208, 223, 230, 233, 280, 282, 285, 292, 300, 302, 303
 David Falconer, 49
 Darrell Hoeman, 99
 Trent Steffler, 87

G.E. Appliances, 178, 195

GMFanuc Robotics Corporation, 34

Bob Gangloff, 20, 26, 29, 30, 31, 32, 37, 43, 57, 71, 74, 75, 83, 84, 85, 86, 95, 105, 106, 111, 113, 116, 121, 124, 125, 137, 231, 237, 247, 256, 267

Ann Garvin, 7, 19, 44, 117, 191, 207, 241, 266, 271

Gateway Assoc., 306

General Motors Corporation, 9, 161

Gerber Products, Inc., 55

Goes Lithographing Company, 38

Gorman Typesetting, Inc., 174

Jeff Greenberg, 120

Photo Courtesy Greyhound Dial Corp., 24

Grumman Corporation, 58

Gulfstream Aerospace Corp., 169

Harley-Davidson, Inc., 142

Hewlett Packard, 45, 144

Linda Henson, 134, 136, 137, 241, 274

John Hopkins Photography, 215

Text Design: Design Five
Cover: The Stock Market
Title Page: Masterfile/Mark Tomalty

Index